The

Clone

Age

The
Clone
Age

Adventures in the New World
of Reproductive Technology

· · · · · · ·

Lori B. Andrews

An Owl Book

Henry Holt and Company · New York

Henry Holt and Company, LLC
Publishers since 1866
115 West 18th Street
New York, New York 10011

Henry Holt® is a registered trademark
of Henry Holt and Company, LLC.

Published in Canada by Fitzhenry & Whiteside Ltd.,
195 Allstate Parkway, Markham, Ontario L3R 4T8.

Library of Congress Cataloging-in-Publication Data
Andrews, Lori. B., 1952–
The clone age: adventures in the new world of
reproductive technology
p. cm.
ISBN 0-8050-6446-X
1. Human reproductive technology—Moral and ethical aspects.
2. Human cloning—Moral and ethical aspects.   3. Human
reproductive technology—Law and legislation.
4. Human cloning—Law and legislation.
RG133.5.A57   1999   98-32099
176—dc21   CIP

Henry Holt books are available for special promotions and
premiums. For details contact: Director, Special Markets.

First published in hardcover in 1999 by
Henry Holt and Company

First Owl Books Edition 2000

Printed in the United States of America

1   3   5   7   9   10   8   6   4   2

For Christopher—

there could never be another you

CONTENTS

∾

The
Clone
Age

Prologue

∽

I am cradled in one of those amazingly large multigadgeted seats in the upper deck of a Singapore Airlines 747. On the personal video monitor attached to my seat, I am watching a cloned Sigourney Weaver swirl a basketball around on her finger, then create an escape hatch by dripping her acid-containing blood on a metal wall.

I take the earphones off in boredom. Watching the movie seems too much like a busman's holiday for me. I am on my way to Dubai, one of the seven sovereigns in the United Arab Emirates. I have been invited by Lieutenant Colonel Abdul Qader Al Khayat of the Dubai Police to advise them about human cloning.

Why, you might ask, does a police chief for an oil sheik need a consultation on cloning? Dubai, like the United States, is harnessing the power to create life in the laboratory. The Dubai government wants my best guess about what will happen once we enter the Clone Age.

For the past twenty years, since my graduation from Yale

Law School in 1978, I have spent much of my life on trips like this—collecting information about genetic and reproductive technologies, and advising judges, heads of state, legislators, medical groups, and others about the societal implications.

On July 25, 1978, the day I took the bar exam, Louise Brown—the first test-tube baby—was born in England. Three years later, I was a speaker in Kiel, West Germany, at the first international meeting of doctors undertaking in vitro fertilization, in which a woman's egg is fertilized with a man's sperm in a test tube (actually a plastic dish) and the resulting embryo implanted back in the woman. In Kiel, Chicago embryologist Dr. Richard Seed proposed splitting an in vitro embryo in half, implanting the first half, and then, in his words, "if it grew up and got into Harvard, we could unfreeze the other half and make its twin." I tried to sort out the family relationships that would result, and the psychological and legal implications of having a twin twenty or thirty years younger.

My work on that problem set the stage for my thinking about cloning now. But in the interim, politicians and judges, reporters and priests have sought my advice on a wide variety of biomedical issues. When a scientist decided to freeze a deceased patient's head with the hope of transplanting it to a healthy body, I was called for an opinion on the rights of severed heads. When an infertility center decided to implant a couple's embryo into a surrogate mother, I was asked to draft the contracts and determine who the legal mother would be. When an elite teaching hospital decided to take sperm from men in comas so that their wives, girlfriends, and, in some cases, even their parents, could create children with the sperm, I was asked about the legal rules that might govern these arrangements and the psychological impact on the resulting child.

By the time I was called by the Dubai police, I was feeling

a little like the Harvey Keitel character in *Pulp Fiction*. I was the cleanup person called in after scientists or doctors had done some strange new thing, the lawyer asked to sort out the rights and responsibilities, the liabilities, and the commercial potential.

Should anything be allowable so long as a lawyer can come up with a scheme to deal with it? Or are there some scientific "advances" that would so change the nature of our society (or so waste money or so damage the participants) that they should be prohibited?

When Dolly the cloned sheep was born, I was asked by the U.S. National Bioethics Advisory Commission for a legal opinion on whether there was any way for President Bill Clinton to stop scientists from doing whatever they please. My 113-page response to that simple question had been posted on the Web by our government, which is how the Dubai Police came to find me.

Before I left the United States, my nine-year-old son, Christopher, and I looked at a fact sheet about the United Arab Emirates, of which Dubai is part:

Political Parties: None
Suffrage: None
Elections: None

My son, imbued with the civic virtue of a fourth grader, wondered how a government could run without those trappings of democracy he was beginning to study.

Yet Dubai has much more in common with the United States than those factors would suggest. It is the most liberal of the Persian Gulf states. While other Gulf nations built their economies on oil, Dubai's strength was always trade and

services. "It is one of the last bastions of anything-goes capital-
ism; sort of an Arab version of Hong Kong," observes travel
writer Gordon Robison. "What opium was to the growth of
Hong Kong in the 19th century, gold was to Dubai in the
1960s. Oil, when it was discovered in 1966, merely contributed
to trade profits and speeded up modernization."

Dubai made its initial fortunes on trade that was, to put
it starkly, illegal. At the end of the eighteenth century, the
British dubbed the area "the Pirate Coast." In 1970, at the peak
of its gold smuggling days, 259 tons of the precious metal
passed through the port of Dubai, mainly on a passage to India.
Today, boats take video players and jeans to Iran and exchange
them for caviar and carpets. "As was the case with gold," notes
Robinson, "all of these goods leave Dubai perfectly legally;
it's the countries at the *other* end of the trade that look on it as
smuggling."

The rulers of Dubai are used to making the impossible possible. Ten years prior to my visit, Dubai had opened the first golf
course in the Gulf with real grass. At other courses in Arab
nations, players lugged a small piece of astroturf to each hole of
the sandy fairways. By the time of my visit, the Dubai Desert
Classic was a well-known stop on the European pro tour. A week
before my visit, Dubai hosted an international horse race, flying
in the winners of the Preakness and the Kentucky Derby. In
1995, the emirate was the site of the Miss World Pageant. Yet
Dubai was not yet a household word in the United States or
Europe. It was itching to be so. In keeping with the history of
making legal in Dubai what is considered illegal elsewhere,
maybe cloning would provide that notoriety.

Dubai, then, would not be unlike the United States, which
serves as an international trade center for reproductive and
genetic technologies, servicing foreigners whose own countries

ban these procedures. People from Europe fly into New York to buy sperm and eggs. A man from China leases a private plane for himself and his concubines to come to California to hire a surrogate mother.

Virtually any reproductive or genetic technology is now available in the United States. Want to get breast cancer genetic testing at a time when professional organizations have said it is premature? Try the Genetics and IVF Institute in Fairfax, Virginia, and Myriad Genetics in Salt Lake City, Utah. Want to choose the sex of your fetus? In Dr. John Stephens's clinic in upstate New York, for $1,200, you can learn the news in the eleventh week of pregnancy—then go to an abortion clinic if the fetus is the "wrong" sex. Want to conceive a child whose genotype is close enough to allow her to be a bone marrow donor to an existing child? The City of Hope National Medical Center Clinic helped Mary and Abe Ayala do just that. Is it the wrong time for a pregnancy, but you want to freeze the fetus in case you later want to bring to life that very baby? Cryogenic Solutions of Houston offers such freezing, even though the technology to "reanimate" such a fetus is not yet available.

And for $3.5 million, Dr. Richard Seed of Chicago says he will clone *you*.

Dubai also shares with the United States a love of science. The framers of the U.S. Constitution discussed the sacred nature of scientific inquiry. The Constitution established a system of patents to promote scientific invention. Historically, scientific theories have been protected because of the great social import the United States places on the value of intellectual freedom. Islamic traditions view scientific inquiry as based on divine warrant, and the Islamic Code of Medical Ethics equates pursuit of knowledge with the worship of God.

U.S. senator Tom Harkin has defended human cloning by

explicitly stating that scientists have the right to research, and that there are not "any appropriate limits to human knowledge. None, whatsoever. . . . To my friends Senator Bond and President Clinton who are saying 'Stop, we can't play God,' I say, 'Fine. Take your ranks alongside Pope Paul V who in 1616 tried to stop Galileo.' " Senator Harkin argues that any governmental ban or limitation on human cloning research is essentially an "attempt to limit human knowledge, which is demeaning to human nature."

But unlike the United States, the Arab countries have not allowed reproductive technologies. Use of artificial insemination is a crime. Dr. Aziz Sachedina, testifying before the U.S. National Bioethics Advisory Commission, explained that the Koran forbids techniques like sperm donation—or cloning a celebrity or relative—because it could disrupt interpersonal relationships, which could "jeopardize the very foundation of human community, namely a religiously and morally regulated spousal and parent-child relationship under the laws of God." In contrast, in the United States, couples have a constitutionally protected right to make reproductive decisions. The extension of this protection to allow couples to use reproductive technology has been hard fought in court cases, including ones I myself litigated. If cloning were viewed in the United States as protected by reproductive freedom, the government would not be able to ban it without proof of compelling harms, such as undue risks to the resulting children.

As the plane begins its descent, I review my instructions about the customs in Dubai. I am not to eat with my left hand, nor point the soles of my feet toward anyone, nor bare my shoulders. After shaking hands, I should briefly touch my heart. My tutor in these matters was Rolph Ehrlich. A retired Air Force colonel, he spent 1979 and 1980 as the executive officer to the commander of all U.S. Forces in the Middle East, headquartered in Saudi Arabia. He was involved in the attempted rescue of the Iranian hostages in Tehran and the recovery of the holy mosque in Mecca. In the 1990s, he left government service to be a principal in an international personal security firm. "We do things that public law enforcement officers are unable or unwilling to do," he explained when we first spoke, the week before my trip. I have left Rolph's phone number with my sister in case I get myself into a jam I can't get out of.

The Singapore Airlines flight attendant interrupts my thinking by handing me a Dubai entry card to fill out. I come to the blank for profession. Harvey Keitel–style fixer doesn't seem quite right. In my purse, I find the faxed visa the Dubai Police sent me and discover that they classify me as a "forensic specialist." That will do.

At Passport Control, I hand over my entry card, passport, and the faxed visa. The guard shakes his head. I will not be allowed to enter without the original of the visa. He points me to a black booth across the waiting room and assures me my visa is probably being held there. The clerk at the booth thumbs through a stack of thirty visas, but mine is not among them. "You may not enter the country," she tells me.

Confined to the waiting area between disembarkation and Passport Control, I am considering my options when a bearded man approaches the visa clerk. He is wearing a kandora, a white floor-length shirt dress, with a white gutra head scarf secured by black head ropes (which, in earlier times, were used to hitch up camels at night). He shoves a piece of white paper toward the clerk and says, "If you see this man, notify me at once."

She nods tentatively in my direction, and he turns around to look at me. His puzzled expression tells me that he wasn't expecting his "forensic specialist" to be a woman.

He recovers quickly and introduces himself as Lieutenant Ahmed Al Bah of the Dubai Police. We are sailing through Customs when I hear a *brmm* noise under his kandora. He reaches inside and pulls out a cellular phone, then speaks quickly in Arabic. Perhaps he is advising someone about my arrival. I cannot tell. My Arabic is limited to *"min fadhlik"* (please), *"shukran"* (thank you), and *"fee aman Allah"* (go with God—a formal good-bye for the end of a journey).

Past Customs, we are greeted by five beautiful grammar school girls in native costumes. The first one tips a large copper carafe toward my hand, dabbing me with scented rose water. Outside the airport, a driver in an identical long white shirt dress waits for us next to a sleek white Cadillac Brougham.

As we drive through Dubai, Lieutenant Al Bah points out the Emirates Training College, fancifully shaped like an airplane. We are on a cloverleaf of a superhighway that was built less than a year ago. Each new building is more whimsical than the last. Concrete and glass have been molded into ephemeral structures that give the impression of billowy white tents. There are high-rise shopping centers everywhere, each increasing in splendor, putting Nieman Marcus and all of Texas to shame with their scale and opulence.

When I enter my hotel room at the five-star Dubai Hilton, I encounter something I have never seen in a hotel room before. A full-sized, carved wooden crib, complete with a Winnie-the-Pooh bedspread.

There is something poignant about this empty crib, which makes me think about the longing of the infertile couples who are desperately trying any means possible to fill such an empty space in their own lives. Previous generations tried fertility rites to enhance reproduction. They believed having sex while the north wind blew would produce a male. Our generation turns to in vitro fertilization, egg donation, frozen embryos, and whatever else the specialist will offer them.

The hotel reminds me of another lesson I learned in my work. People don't just want children, but children with more advantages than they had. There is a striking poster in the window of Damas, the Dubai Hilton's jewelry store. The forty-five-year-old mother in the poster is completely covered with the traditional black Arab garment, which reveals only her eyes. In stunning contrast, her gorgeous twenty-five-year-old daughter appears to be completely naked. (She is photographed from the top of her long cascading dark hair to just above the nipples.) The mother is fastening an ornate gold necklace around the daughter's elegant bare neck. The message seems to be: Give my child the chances I did not get.

With parents' quests to better the life of their children, it seems only natural that couples would want access to egg and sperm from high-IQ donors, to genetic engineering to protect their children from disease, and to human cloning as the ultimate chance to provide their progeny with favored traits. But should those seemingly understandable desires be satisfied?

As I lay out my papers on my hotel desk, I look over a map of the United Arab Emirates printed off the Web. The United Arab Emirates is triangular, with one leg bordering the Persian Gulf. The other two legs, next to Saudi Arabia and Oman, are marked by the words "No defined boundary."

That is how I feel about my work in reproductive technology and genetics.

No defined boundary.

1

Creating Life in the Lab

∾

The year is 1985. The world's television sets are tuned to Bob Geldof's Live Aid concert at Wembley Stadium in London, but I am in another part of England. My taxi is motoring up a long, curvy driveway until it reaches a clearing. Then the Jacobean manor house, Bourn Hall, comes in view, perfectly set in the middle of a rolling green lawn, a herd of sheep to the side. Built in 1607 as the seat of Earl de la Warr, the mansion betrays nothing about its present purpose. But a few miles back, the small local store has added a new line of products. Amid the hand-knit sweaters for which this part of England is famous, there is a display of baby bottles, bibs, and gowns.

The sitting room at Bourn Hall looks like a United Nations of women. Some are reading, others knitting, many are talking. Their bodies seem bent with grief, but there is hope in their eyes, like cripples approaching Lourdes or the Catedral de Guadalupe. They are barren women. They have come here from countries around the globe so that Robert Edwards can create a baby for them. So many have come that Edwards has erected trailers in the back of the mansion to house these women in waiting.

I am thirty-two, about the same age as many of these women, married for four years, and childless by choice. My biological clock hasn't started ticking loudly enough to raise concerns, but I can empathize with the stories they tell me of the enormous lengths they are going to in order to create a child. I can imagine their disbelief when they stopped the Pill or threw away their diaphragms and first learned that they could not conceive.

They are part of a generation of women—my generation— for whom a different set of doors had been opened. We were young at the moment that everyone else was trying to be. We were in the first group of women admitted to previously male colleges, clubs, professions. Abortion was legalized just when we first needed it.

But it seems that a strange thing had happened while life was saying yes to us; our bodies increasingly were saying no. Women who postponed childbearing to pursue educational and career opportunities suffered the natural decline in fertility that occurs with age. Women who exercised their right to contraception, in ways such as using the Dalkon Shield intrauterine device, sometimes found their fertility permanently compromised. For many, the only hope was to use Edwards's magical new panacea, in vitro fertilization.

Edwards, an embryologist at Cambridge, had spent two decades trying to fertilize women's eggs outside their bodies. In his early experiments, he used his own sperm and that of male lab techs in attempts to fertilize eggs recovered during hysterectomies. Later, while on a research fellowship in Chapel Hill, North Carolina, he invented a porous little cage to contain egg and sperm and persuaded women to volunteer to wear them in their uteruses.

The British Medical Research Council denied funds to Edwards and his obstetrician collaborator, Patrick Steptoe, because of concerns about the ethics of their work. At a 1971 meeting in Washington, D.C., James Watson, the codiscoverer of DNA, chastised Robert Edwards. "You can only go ahead with your work if you accept the necessity of infanticide. There are going to be a lot of mistakes."

Edwards's money came primarily from generous individuals (predominantly Americans), but also from a Ford Foundation grant for studying in vitro fertilization to aid in the development of contraceptives. Ironically, it was this grant for research on *preventing* pregnancy that led to a successful pregnancy in 1978.

After about one hundred attempts in other women, Edwards fertilized Lesley Brown's egg in a petri dish with the sperm of her husband, John. Three days later, in the middle of the night, Edwards gingerly implanted the embryo back into Leslie and the pregnancy began. Newspapers, magazines, and television stations treated the impending birth of a baby to the Browns as an extraordinary event. When the pregnant Mrs. Brown was hospitalized, press people dressed up as boilermakers, plumbers, and window cleaners to sneak into the hospital to try to interview or photograph her. One exasperated hospital representative complained, "It seems as if you move anything, there is a reporter behind it." Someone—probably a reporter—called in a bomb scare to the maternity building, most likely in the hopes of getting a glimpse of Lesley Brown. Every woman had to be evacuated, including those who were in labor or who had surgery earlier that day.

The newspapers even surreptitiously obtained—and

printed—reports of confidential medical tests that had been performed on Lesley to monitor the development of the child within her. One day the shock of reading a newspaper report drastically raised Lesley's blood pressure. She asked her obstetrician, Patrick Steptoe, in tears, "Is it true that my baby nearly died yesterday? Is it true? Will it happen again?"

Steptoe assured her all was well, and, true to his word, on July 25, 1978, she delivered a little girl, Louise Brown, the world's first in vitro fertilization baby. The media breathlessly reported the birth of Louise Brown with a fervor not seen since the first moon landing.

. . .

Edwards's wife, scientist Ruth Fowler, was no stranger to controversy. Her grandfather, Ernest Rutherford, had split the atom. *The same genes as Rutherford,* Edwards thought when he first met Ruth, *my God!* But even that precedent could not prepare him for the furor that awaited.

The church and Parliament went wild. Immoral, they said. Unnatural. Playing God. Even scientists were on edge. Some didn't believe it had occurred and demanded a peer-reviewed publication. Others criticized Edwards for going ahead before the procedure had been done in higher animals. No one had even tried it in a chimp! Still others worried about the limits that would be put in place on other scientists if one of Edwards's creations turned out to be defective and the Parliament, in horror, clamped down on all avant-garde research.

The procedure of in vitro fertilization itself was not just some overnight sensation, having been suggested as early as 1937 in an editorial in the *New England Journal of Medicine* entitled "Conception in a Watch Glass." Edwards began publishing reports of his own fertilization attempts beginning in the late 1960s.

In 1973, Floridians Doris and John Del Zio became the first couple in the United States to attempt IVF. Doris's New York City infertility specialist, Dr. William Sweeney, had tried three times to repair her fallopian tubes without success. Sweeney felt additional surgery would be of no benefit, so he asked Dr. Landrum Shettles of Manhattan's Columbia Presbyterian Medical Center to assist him in a radically new procedure.

The hospital was on East 70th Street in New York, while Shettles's laboratory was on West 168th Street. On September 12, 1973, Sweeney surgically removed an egg from Doris Del Zio, putting it in a sterile container with protective packing. John Del Zio then carried the package across town in a taxi to Shettles's Columbia Presbyterian laboratory. There, after the addition of the husband's sperm, the culture medium containing the egg was placed in an incubator to develop for three days before an attempt at implantation.

The day after removing the egg, Shettles was ordered to the office of his boss, Dr. Raymond Vande Wiele, the chairman of the Department of Obstetrics and Gynecology at Columbia Presbyterian Hospital and Medical Center. Vande Wiele was furious that Shettles had attempted this sort of experiment without seeking the institution's permission. Vande Wiele argued that it was unethical and immoral to carry on this work. Later, he would claim that he had been worried that the contents of the petri dish were contaminated and that he did not want to see Del Zio risk infection or even death if they were implanted in her uterus.

As the two men spoke, Shettles noticed the Del Zio container sitting on Vande Wiele's desk. The department chairman had removed it from the incubator. He had also opened it, thus dashing completely the couple's hopes of giving birth to a child through in vitro fertilization.

At that moment, Doris Del Zio was still hospitalized, recovering from the surgery. When the news of the chairman's action reached her, she sank into a profound depression. Sweeney had told her that her reproductive organs were so badly damaged he could not attempt to remove another egg, so she had lost her one chance to become a mother. A year later, she and her husband filed suit against Columbia Presbyterian Medical Center, Columbia University, and Raymond Vande Wiele.

In 1974, when the suit was filed, there seemed to be little chance of its success. The whole process of IVF seemed like something out of a science fiction story, so it would be difficult to convince a jury that the Del Zios had indeed lost a realistic chance to become parents. The wheels of justice grind slowly, however, and the trial did not begin until July 17, 1978.

Nine days later, headlines screamed the news of Louise Brown's birth in England. A defense attorney tried to prevent the jury from being influenced by the success across the ocean. He said scornfully that Shettles's procedure was as distinct from the English technique as "a Model T Ford is from a Porsche."

The lawyers were at a loss to define the harm that had been caused to Mrs. Del Zio. It seemed crass to think of the mixture of egg and sperm as "property." She thought of it as her potential child, but the law offered no clear protections. Not knowing what to do with the case, the jury awarded Mr. Del Zio $3—far less than he could have gotten selling sperm as an anonymous donor. Mrs. Del Zio had so obviously suffered, though, that the jurors awarded her $50,000 for her mental anguish. The jury decided that the defendant's conduct was so extreme, outrageous, and shocking that it exceeded all reasonable bounds of decency.

By the time I visited Robert Edwards in 1985, more than 1,300 doctors and scientists were quietly working on in vitro fertilization and similar reproductive technologies. Even Dr. Raymond Vande Wiele had succumbed to the allure of IVF and opened his own IVF clinic at Columbia. In the Clone Age, it would be possible for a child to have five parents: a sperm donor, egg donor, surrogate mother, and the couple who intended to raise the child.

• • •

The reason that procreation started moving so quickly from the marriage bed into the scientific laboratory in the 1970s was that infertility had become a problem for an increasing number of people. According to a study by Dr. Ralph Dougherty, professor of chemistry at Florida State University in Tallahassee, the sperm count of American males has fallen dramatically. In 1929 the median count was 90 million sperm per cubic centimeter; in 1979 it was 60 million. Men born in the 1970s had 25 percent fewer sperm than men born in the 1950s. Possibly because of environmental pollutants, nearly one quarter of all men had counts so low that some scientists consider them to be "functionally sterile."

Women, too, were having problems. Young women had been having sex with a variety of partners and sustaining untreated low-level gynecological infections, which damaged the reproductive system. Before the 1960s and the development of the Pill, the main contraceptive devices—the condom and diaphragm—had protected against infection. According to Dr. Robert T. Francoeur, professor of human sexuality and embryology at Fairleigh Dickinson University in Rutherford, New

Jersey, as many as one in four women between the ages of twenty and thirty-five are now infertile.

I originally met Robert Edwards in 1980 at the first international meeting on in vitro fertilization, a conference in Kiel, West Germany, where I was speaking on the legal implications of surrogate motherhood. At that time, Louise Brown was the only confirmed in vitro birth. Edwards's next patient to become pregnant miscarried because the fetus had sixty-nine chromosomes instead of the normal forty-six. The patient after that miscarried at twenty and a half weeks, before the baby was mature enough to survive. Another implantation resulted in a tubal pregnancy that had to be surgically removed. Yet the hundreds of doctors at the conference were ready to plunge forward with a panoply of brave new ways to create the next generation.

Since so little was known about the best way to fertilize an egg in vitro, two brothers, Dr. Randolph Seed, a physician, and his brother, Dr. Richard Seed, a veterinarian, thought that they would rely more closely on nature. They proposed to use a female volunteer as a human petri dish. They called it in vivo fertilization, rather than in vitro.

When Richard Seed approached the podium in Kiel, I marveled at the irony of his name. He was wild-eyed, confident. He described a procedure developed to transfer embryos from prize cows to ordinary cattle as a way to upgrade the herd. He proposed using that same approach to flush embryos out of fertile women and put them into the infertile.

In artificial embryonation (AE), sperm from the husband of an infertile woman would be used to inseminate a fertile woman. Then, four to five days after fertilization, the embryo would be flushed out so it could be implanted in the infertile woman, as in in vitro fertilization.

In addition to artificial embryonation, the Seeds planned to offer embryo adoption (EA). "The technology is the same as an artificial embryonation, but donor semen is used instead of semen from the recipient's husband," said Dr. Richard Seed. "In this case, the embryo has genes from neither of these 'adoptive' parents, even though the 'adoptive' mother will bear the child."

Seed told the Kiel conference how the idea came about. "It's amusing I didn't think of this myself," he recalled. "A patient called up and said he had been trying to adopt a child for five years, and asked, 'Why can't I adopt an embryo?' A few months later I called him back, and that started the whole thing."

Richard Seed confidently cited seven years of experience doing embryo transplants in cattle to justify his optimism that his procedures would succeed. "Ten thousand transfers have been done with cattle," he said, "and there are no excess abnormalities." He later added, "I see no reason why it shouldn't work with humans."

Several doctors at the Kiel conference attacked Seed's attempts to draw conclusions about human beings from experiments done with animals. Dr. Martin Quigley said that rat eggs did not fertilize in vitro under the same circumstances in which mouse eggs were routinely fertilized. If one could not generalize in animals so closely related, how could one hope "to establish standards for human in vitro fertilization from laboratory experiments with animals?"

Seed believed that neither AE nor EA would involve any legal procedures. "We feel the child delivered to the previously infertile mother will be her child legally," he explained, "and her previously infertile husband will be the legal father, with no necessity for actual adoption proceedings." This view was in keeping with most of the artificial insemination cases, which say that the woman who bears the child is the natural mother. But considering that in artificial insemination cases a child is not only borne by the mother but also conceived of her egg, who knew how a judge would feel if one of Seed's volunteers, who donated the egg—and thus half the genes—for the child, sued to have visitation rights?

The idea was not as far-fetched as it may sound. In England, a British doctor had planned to help an infertile woman have children by doing an ovary transplant. The surgery was stopped, however, when British authorities informed the doctor that, although this patient could then conceive and carry a child, any offspring that might result would be considered illegitimate. In such a case the legal mother might actually turn out to be the woman who gave up the ovary and thus donated her eggs.

The creation of embryos—whether in vitro, as Edwards did, or in vivo, as Seed did—might lead to procreation by design, with parents picking the traits they wanted in their offspring. E. S. E. Hafez predicted that it would give rise to ova and embryos being "banked" and sold for commercial purposes. He said that it would be possible to have an embryo supermarket. According to Hafez, the purchaser would be able to select an embryo knowing in advance the "color of the baby's eyes and hair, its sex, its probable size, . . . maturity, and its probable IQ."

I was the only American lawyer on the Kiel program, and as I listened to each new speaker, I thought about the legal implications of their work. If an unmarried woman used donor sperm, could she sue the donor for child support? Could the resulting child, like adoptees in some countries, learn the identity of her donor father?

I also relished the irony of the Seeds' address. Back in my hometown of Chicago, their medical office was in Water Tower Place, a shopping center on Michigan Avenue that houses some of the most exclusive shops in the world. There you can buy everything from eighteenth-century Chinese screens to remote-control robots. Now, in a suite marked "Reproduction and Fertility Clinic," you could buy human embryos.

• • •

For years after the Kiel meeting, the doctors in attendance called with their latest legal conundrums. One had inseminated a woman's egg with her husband's sperm to be implanted in the husband's sister as a surrogate. As he was about to put the embryo in the sister, he stopped suddenly and rushed to the phone. If he put the man's embryo in his sister, would that violate his state's ban on incest? If he decided not to go through with the implantation and terminated the embryo instead, could he be found guilty of murder?

The law addresses any new question by relying on precedent. When cars were introduced, the wisdom from cases dealing with horses and buggies governed. But, as the Del Zio case demonstrated, it was difficult to find precedents to deal with human embryos. Were they property or people? Where could we turn to learn if they could be frozen—or "enhanced" through genetic manipulation?

The doctors called me, they said, because the attorneys for their clinics respond to each of their questions by saying, "There is no law in this state that covers that." Which is not much comfort for a doctor with an embryo in a catheter who is trying to decide what to do.

In 1969, when Edwards first published an article about fertilizing human eggs in *Nature*, William Breckon, the science correspondent for the *Times*, wrote, "Ultimately we could have the know-how to breed these groups of human beings—called 'clones' after the Greek word for a throng—to produce a cohort of super-astronauts or dustmen, soldiers or senators, each with identical physical and mental characteristics most suited to do the job they have to do."

Some scientists wanted to try crossing species, ape egg and human sperm, or vice versa. In the late 1960s, Chinese surgeon Ji Yongxiang reportedly fertilized a chimpanzee with human sperm to create a "near human ape" to perform simple tasks. The pregnancy was terminated when rioters destroyed his lab.

An article in a law journal has raised the question whether hybrids would have the same legal rights as humans. If the hybrid is half human, would it have the right to vote? Half a vote? What about its offspring, who might be only a quarter human?

Many people wanted to ban IVF to prevent us from careening down such a slippery slope. But the Chinese attempt to merge human and ape did not rely on IVF. Even if IVF with its laparoscopy component were banned, scientists could use eggs from the ovaries of women whose ovaries had been removed in the course of a complete hysterectomy. What would the opponents then propose—that hysterectomy be banned on the off chance that a scientist might decide to breed a genius hybrid with the egg?

• • •

All around the world, in vitro fertilization clinics were coming under attack. Because not all of the embryos resulted in babies, right-to-life groups protested. In Australia, priests staged a hunger strike in the lobby of Alex Lopata's clinics. His patients had to walk past this human barrier whenever they wanted to see him. The strain was palpable. In one instance, a slight young woman calmly walked into the heat of the protest. She looked a priest directly in the eye and asked quietly, "Why don't you want me to have this baby?"

Shortly after Louise Brown's birth, right-to-life advocates persuaded Illinois lawmakers to pass an unusual law to deter doctors from doing IVF. The law said that any physician who fertilized an egg in vitro had custody of the resulting embryo and would be subject to an 1877 child abuse law. Doctors in Illinois were appropriately intimidated. They knew what it meant to provide an existing child with the food, clothing, and shelter necessary to avoid a finding of abuse, but what did it mean in terms of an eight-cell embryo? Could a prosecutor indict a doctor each time an embryo failed to develop into a child—on the grounds that the doctor should have "fed" it a more "nutritious" petri dish mixture? Would the doctor be guilty of homicide if he or she discarded an embryo that was not dividing properly?

Another land mine was that the law granted custody to the doctor but never arranged for the parents to *regain* custody! If such a law had been in effect in England, Lesley and John Brown would have had no legal claim to Louise; she would have belonged to Edwards and Steptoe.

Lois Lipton of the American Civil Liberties Union was approached by a woman who wanted to challenge the law. Her fallopian tubes had been damaged nineteen years earlier when her appendix ruptured. She underwent a number of corrective operations on her tubes, but the original damage was too great. Her doctor, Aaron Lifchez, had advised her that IVF was her only available method of conception. Lois called me and asked if I'd be willing to help in the case.

I knew that the chances of the woman being accepted into an IVF program in another state were very slim. Clinics at that time had an age thirty-five cutoff. She was thirty-four years old when she filed suit and would be too old to qualify for some programs if she were on a waiting list for more than a year. Besides, she had confidence in her own doctor; the only reason he was unwilling to offer in vitro fertilization was his fear of being prosecuted for child abuse.

Two other women joined the team: Frances J. Krasnow of the Chicago law firm of Fohrman, Lurie, Sklar & Simon, Ltd., and ACLU attorney Colleen Connell. We read and reread the contraceptive and abortion decisions, taking solace in the U.S. Supreme Court's language: "If the right of privacy means anything, it is the right of the individual, married or single, to be free of unwarranted governmental intrusion into matters so fundamentally affecting a person as the decision whether to bear or beget a child."

There was no question that a married couple has a right to determine whether and when to bear a child through intercourse. But did a couple also have the right to decide *how* they would like to bear a child?

"While the legislature apparently views *in vitro* fertilization as a crime," we wrote in our brief, "to many childless couples it is seen as a possible miracle. . . . Procreation is universally recognized by every culture and religion as a fundamental element of the institution of marriage. For many married couples it is the essence of family. The desire to produce one's own offspring is, for most couples, as primary as the need to eat or sleep." We argued that the Illinois law interfering with that decision should be declared unconstitutional.

After we filed suit, the Illinois attorney general and the state's attorney for Cook County rendered an opinion that any

physician who undertook the procedure would not violate the
law as long as the physician refrained "from wilfully endangering or injuring" the embryo during the preimplantation period.
They said that if an embryo was terminated because it was
defective, that action would be interpreted as a lawful pregnancy termination on the part of the physician. The legislature
passed a law allowing in vitro fertilization. But then it made any
research involving embryos a crime.

Once again, we were back in court.

For our brief, I scoured hundreds of medical journals to
demonstrate how a ban on embryo research would interfere with
infertile women's chances to create a child. In vitro fertilization
and reproductive technologies were so new that they were
inherently experimental. Article after article described how
doctors were varying every aspect of the procedure—trying
new culture media, changing the shape of the petri dish—in
order to improve the success rate of IVF. Randolph and Richard
Seed had teamed up with a group of doctors at UCLA to offer
artificial embryonation, and yet that was clearly banned by the
Illinois law. So was freezing an embryo for later implantation
or undertaking genetic screening on the embryo. This seemed
to me clearly to violate the woman's right "to bear and beget
a child."

And the law had another fatal flaw. It violated doctors' constitutional rights by failing to give them notice about what constitutes criminal behavior. Embryo experimentation was
unlawful, but *experimentation* was never defined. The medical
articles I had consulted offered differing—and contrasting—
definitions of what was an experiment. Some argued a procedure
was not an experiment, even if it had never been done before, so
long as the doctor "intended" it to be therapeutic. Other articles
said that even a standard procedure was an experiment when it

was tried on a new patient. Federal regulations considered a procedure experimental only if it was part of a larger study.

The transcript of the Illinois legislature showed that even the men passing the bill had no idea what it meant. Some thought it made prenatal screening through chorionic villi sampling illegal; others thought it did not. It was a fundamental principle of U.S. law that if a state created a new crime, it had to be clear about what actions were forbidden. We could make a strong argument that the law should be "void for vagueness."

We didn't want to see hundreds of human embryos lined up in rows with researchers testing drugs or cosmetics on them, but the Illinois ban was just too broad, preventing women from using legitimate fertility procedures and genetic testing.

The right-to-life groups were up in arms. Their protestors came to court to hiss when we went by. My co-counsel, Lois Lipton, was very pregnant herself at one court hearing. A right-to-life protestor in the back row uttered a death threat against her. There was a right-to-life hotline where you could hear recorded death threats against Lois anytime, night or day.

I was working peacefully at my office one day when a burly, angry man barged in and served me with a legal complaint. He was bringing a class action on behalf of all the unborn children in Illinois. He claimed to represent all *unconceived* children. He was representing my eggs—in a suit against me.

The right-to-life groups were not our only critics. Even some feminists opposed in vitro fertilization. "What is the real meaning of a woman's 'consent' in a society in which men as a social group control not just the choices open to women but also women's *motivation* to choose?" the feminist Gena Corea asked. She said that Lesley Brown probably chose motherhood only to escape her boring job in a cheese factory.

Weeks, months, more than a year passed before we heard

from the court. One day I got a call from the young lawyer who served as the judge's clerk. By the following day he wanted copies of the dozens of medical articles I had included in the brief. I explained how I had returned those journals to medical libraries across the city and asked what would happen if I couldn't produce them in time.

"You'll be in contempt of court," he said.

The federal judge, Ann Williams, was a woman—just thirty-five years old when she was appointed to the federal bench in 1985. Her opinion read: "It takes no great leap of logic to see that within the cluster of constitutionally protected choices that includes the right to have access to contraceptives, there must be included within that cluster the right to submit to a medical procedure that may bring about, rather than prevent, pregnancy."

She threw out the whole law as unconstitutional. She also dismissed the man who claimed to be representing the state's unconceived children.

When I read about the case the next day in an Associated Press article, the important issues of women's reproductive rights were nowhere to be seen. The story highlighted the fact that researchers could now use human embryos however they wished—including slicing them up to inject as a treatment for Parkinson's patients. It was my first legal brush with the principle of unintended consequences.

●   ●   ●

When Antonie van Leeuwenhoek first viewed sperm under a microscope in the 1670s, the biological groundwork was set for legal questions about the manipulation of reproduction. But it was not until the clinical application of in vitro fertilization three centuries later that the question took on a pressing significance. Once the embryo was isolated in the petri dish, it could be used to create a child for the progenitors, it could be donated to another couple, it could be genetically manipulated, or it could be used for other research purposes. As Clifford Grobstein has pointed out, the work that Robert Edwards was doing was "moving from unconscious cultural determination of human biological progression to a degree of conscious self-determination."

At Bourn Hall, I was surrounded by the trappings of the past. The elegant sitting rooms and burnished wood banisters seemed an appropriate setting for a play of English manners. But as Edwards ushered me into his embryo laboratory, I was transported ahead in time, with the gleaming steel equipment and slightly antiseptic smell. This was why the law had not kept up with the changes in reproductive medicine. If a lawyer from a hundred years ago were set down in a modern courtroom, he— and, of course, it would be a *he*—would feel completely at home. But if a doctor from the 1800s were to find himself among Edwards's liquid nitrogen tanks, laparoscopy equipment, and ultrasound, he wouldn't have the foggiest idea of what to do.

The laboratory gleam silenced me, commanding the respect that, in early generations, one would have given a church. Here, Edwards explained to me, he would be creating a new genesis. Edwards desired not only to facilitate fertilization, but to improve upon it. He was angry that the British government was restricting embryo research, since he wanted to undertake genetic interventions on embryos. "The dogma that has entered biology from Christian sources has done nothing but harm," he told me.

I left Bourn Hall shortly before dusk. The women there were readying themselves to have their embryos implanted. Edwards liked to do it in the evening, perhaps out of superstition, because Lesley Brown had achieved her pregnancy when the implantation occurred at night. The same thing had happened, more recently, when he provided Louise Brown with a baby sister, Natalie. But perhaps biology played a part as well since the hormones necessary for sustaining a pregnancy are at their highest level at night. The adrenal glands, controlling the body's day-to-day activity, are least active, slowing down the body's processes and creating a more restful home for the embryo.

As they readied themselves for their evening implantation, some of the women could not forgo the vestiges of romance. One carefully combed her long blond hair, applied makeup, and slipped into a magenta velvet gown.

2

Monitoring Medicine

∽

D r. Howard Jones and his wife, Dr. Georgeanna Seeger Jones, were the distinguished father and mother of in vitro fertilization in the United States. Years before Louise Brown, Howard Jones had been Robert Edwards's mentor while Edwards was on a U.S. fellowship. Professors emeritus at Johns Hopkins School of Medicine, they had retired to modest posts at the Eastern Virginia Medical School. As they were moving into their new home there, a local reporter called them for a comment on the birth of Louise Brown. How much money would it take, asked the reporter, for them to start an in vitro program in Virginia? Georgeanna rattled a figure off the top of her head: "$5,000." A few days later, a check in that amount arrived in the mail from a former patient.

Three years later, the Joneses announced the first birth of an in vitro child in the United States, a baby girl named Elizabeth Carr, and other researchers raced to keep up with them. Dozens of in vitro clinics subsequently opened their doors, with no state or federal guidelines to regulate them.

In 1974, Dr. Pierre Soupart, a professor at Vanderbilt University, had submitted an application for funding to the National Institutes of Health (NIH). Two years earlier, in 1972, he had been the first American scientist to prove he could fertilize a human egg in vitro. In his 1974 funding request, he proposed a three-year, $375,000 study to determine the safety of in vitro fertilization. With the cooperation of women who were undergoing gynecological surgery for other reasons and would donate eggs, Soupart planned to fertilize 450 eggs, grow them around six days, and study their chromosomes to see whether the in vitro fertilization process caused any chromosomal abnormality. Soupart's proposal was for research only; he did not intend to implant any fertilized embryos.

In the spring of 1975, Soupart learned that a scientific review committee at NIH had decided to grant him the money he needed to complete his study. There was one hitch, though. A brand-new law now required that any IVF proposal had to be reviewed by the Ethics Advisory Board (EAB) of the Department of Health, Education and Welfare. But in a strange catch-22, the board had not yet been appointed. In fact, it took another two years for the government to appoint its members. It was not until 1978 that the board began its deliberation of Soupart's proposal.

The fifteen-member board, which included physicians, medical ethicists, lawyers, a psychiatrist, and civil leaders, worked from May 1978 to May 1979 gathering information and preparing its report. Board members held hearings across the nation, taking testimony from medical and scientific experts, theologians, people with infertility problems, concerned individuals, professional organizations, and various special interest groups. They received 2,000 pieces of correspondence to guide them in their quest. They hired consultants to prepare reports on the massive literature that existed about IVF.

The EAB's task took on an added importance when, two months after it began its analysis of the IVF issues, baby Louise was born in England. Ironically, if Soupart had been given the funding when he initially asked for it, the first test-tube baby might have been American. The board concluded that the National Institutes of Health should fund IVF, just as it funded any other medical procedure.

But the secretary of Health, Education and Welfare (as well as all subsequent secretaries of Health and Human Services) was too intimidated to fund IVF. Pierre Soupart died on June 10, 1981, without receiving the grant he had applied for. And IVF was left to develop in a gangly, unsightly way, without federal supervision.

Most other medical procedures in the United States have followed a more orderly path. NIH would fund an initial study in which only a few medical research centers undertook the procedure, designed ways to improve it, and assessed any potential risks before the procedure became widely used.

For IVF, there were no federal research funds; nor did health insurance cover it. The growing array of IVF clinics competed commercially for patients and used women themselves as guinea pigs. In vitro was done on women in 1978, but not on baboons until 1979 and chimps until 1983. This led embryologist Don Wolf to quip that perhaps women were serving as the model for nonhuman primates.

Wolf's collaborator, Dr. Martin Quigley, from the University of Texas Health Science Center at Houston, who established the second in vitro fertilization clinic in the United States, commented: "Even when animal studies have been done with this technique, it has told us very little about the abnormalities IVF can produce. This is because most animal studies are concerned with achieving fertilization and do not go on to transplant the fertilized embryo in a host mother to see what type of offspring might be created." However, defects in offspring, including deformities of the pelvis and eyes, *had* shown up in some of the research where animal embryos actually did progress to birth, leading to questions about whether humans would have similar deformities.

Since the federal government wasn't regulating in vitro fertilization, the medical organization of infertility doctors, the American Fertility Society (AFS), decided to set up its own committee, with Howard Jones as the chair, and with me as a member. One concern that I had on the AFS Ethics Committee was ensuring that women got honest and complete information about the risks of IVF—and the low chances of success. When I had attended the meeting in Kiel, an estimated 20,000 women had sought IVF, but only three babies had been born. Yet doctors were telling patients their chance of success was 1 in 10, not 1 in 6,666. A 1985 survey of 169 in vitro clinics in the United States found that half had never had a live birth—not a single birth! Yet they were not telling women that fact.

Even some of the clinics that had babies were misleading women in their brochures. Some doctors claimed to have "trained with Robert Edwards," when all they had done was pay him a visit. I was able to get language into the AFS Ethics Code saying that the doctor must provide the women with his *own* success rate, not just the industry average. Later, Virginia adopted a law with the same requirement, and, much later, the federal government began to collect—and publish—individual success rates of IVF clinics.

I also worried about women who underwent hormonal stimulation and surgery for their own IVF or to donate eggs. While semen is usually collected in "masturbatoriums" —softly lit rooms filled with *Playboy* and *Penthouse* magazines, collecting an egg in the early years of IVF necessitated surgery and thus presented considerable risk. In Kiel, I had learned that a woman had died during a laparoscopy (a procedure that is often used in the collection of an egg). Other women had been killed when their ovaries literally exploded from too much hormonal stimulation.

At a Florida IVF clinic, a woman and her husband underwent IVF. When twins were born, blood testing revealed a mix-up with the sperm. The babies were genetically the wife's, but not the husband's. The couple's relationship deteriorated when the dad discovered that his son, his namesake, was not "his" at all. The situation was complicated by the fact that the wife was white, the husband was black, and—because a white man's sperm had been used—the children obviously didn't resemble the husband.

In a clinic in the Netherlands, a technician apparently did not clean a pipette between uses. One of the wife's eggs was fertilized with her husband's sperm, but apparently the pipette still contained another man's sperm, which fertilized a second egg. She gave birth to fraternal twins of two different races, only one of which was her husband's.

Both of these mishaps were discovered because children of an unexpected race were born. Given that over 99 percent of IVF patients and their husbands are white, who knows how many mix-ups may have occurred without couples ever realizing it.

The blatant commercialization in this field troubled me, too, and the resulting high cost of services. Richard Seed had lured venture capitalist Lawrence Sucsy into a partnership to form clinics to apply the cow-flushing technique—artificial embryonation—to women. In 1984, Sucsy told *Fortune* that the goal of the resulting company, Fertility and Genetics Research (FGR), was to set up a chain of fertility clinics nationwide, possibly franchised through medical groups, and that he was trying to enlist hospitals as moneymaking partners in the venture. Sucsy stated that FGR planned to recruit hospitals that would then give FGR a long-term, low-interest loan of about $1 million in exchange for part ownership in a clinic. He saw a huge market in fertility services because of the high incidence of infertility in the United States, estimating that from 30,000 to 50,000 women per year would be candidates for embryo transfers. The company intended to treat them at an average cost of between $4,000 to $7,000 per attempt. Suscy said, "Volume is nothing, margin is everything."

More than 2,000 women enlisted in the first few months. Richard Seed told the *New York Times* that the procedure would cost the couples "as much as a new car. It should be painful to them, otherwise it isn't worth our effort to work with them for a year or two."

As a physicist and cattle breeder, Richard Seed had no right to practice medicine, so he expanded his team to include John Buster, an infertility specialist at UCLA Harbor Medical Center. Buster had been working in the field of embryo transfer techniques since 1979. He had contacted the National Institutes of Health in 1980 but was told that no money was available for the development of technology relating to embryo transfer. He then contacted Seed, and in 1982, FGR invested $500,000 in Buster's UCLA project.

When I visited the UCLA center, I learned that while the doctors were making thousands of dollars per attempt, the women who were providing the eggs and allowing themselves to be inseminated and flushed were only making $50 (with a $200 bonus if a fertilized egg was recovered). Richard Seed explained it this way: "This is a typical free-market activity. We have investors expecting to obtain a return on their money."

I talked to one of Buster's egg donors, Cyndy Imhof, when we were both testifying before Congress in 1984, and was taken aback by her eugenics zeal. When Cyndy Imhof was asked by the Harbor-UCLA Hospital team why she wanted to be a donor, her reply was just to hold up a picture of her seven-year-old son. "I come from an extremely healthy background without cancer, diabetics, disfigurements, or catastrophic diseases," she explained to me. "I couldn't see why I should just waste these wonderful genes I was blessed with."

Once in the program, Cyndy found that her friends and family were thrilled with her role in creating a baby for a childless couple. "My son is probably the most enthusiastic," she asserted. "One month he asked if 'we' were going to make a baby for another lady. When I answered no because there wasn't a match, he stood there and said, 'But Mommie, you're just wasting your egg!' It could not have been said any better. Why should any woman waste her eggs when there are so many women who desperately want children and cannot conceive?"

By 1987, Richard Seed said that FGR intended to operate up to fifty more embryo transfer clinics nationwide within the next five to seven years. By 1987, FGR was operating with $8 million of private capital. It went public with the NASDAQ listing of BABY.

Then John Buster announced he planned to patent not only the catheter that was used in artificial embryonation but also the procedure itself. The latter was unheard of. What if someone had patented appendectomies? If my doctor didn't want to pay a royalty, would he use a less safe procedure—maybe taking my appendix out through my chest, rather than my abdomen?

Buster bragged, "I think it is going to become one of the key landmark patents that will allow a corporate structure to build a whole health care system surrounding infertility." Many physicians worried that the FGR's patenting was setting up a disturbing precedent. The company intended only to grant licenses to a few groups of reproductive specialists.

Enough doctors were up in arms that they brought the matter to the AFS Ethics Committee. Dr. Richard Marrs, an in vitro specialist on the committee and professor at UCLA Medical Center, was opposed to FGR's patent, saying, "It limits the clinician's ability to treat a patient."

Committee member Celso Ramon Garcia, an infertility specialist from Philadelphia, marveled at how low his field had sunk. Garcia had started his career with John Rock, one of the inventors of the birth control pill. Garcia had been offered stock in the company that made the Pill—enough stock to have made him a millionaire—but he had refused it, feeling it would create a conflict of interest since his research involved assessing the risks and benefits of the drug. Obviously, things had changed.

The AFS Ethics Committee condemned the patenting of the procedure as unethical, but its pronouncement had little effect on the U.S. Patent Office. Unlike in Europe, where patents can be denied on moral and public policy grounds, the United States will issue a patent on anything that is "novel, non-obvious, and useful" no matter what protests are made. In 1987, it issued patent number 5,005,583 on the embryo transfer procedure.

Infertility was now big business, and Sucsy envisioned an

unlimited market of customers. Sucsy also spoke of embryo transfer being used to replace prenatal screening through amniocentesis; he theorized that an embryo could be removed from its mother's womb for genetic testing and then returned to her.

Ultimately, the embryo transfer technique did not develop in the same exponential way as did IVF. Other doctors were unwilling to pay the licensing fees, so they did not experiment with the artificial embryonation procedure to improve it. The UCLA team and the Ovum Transfer Center in Milan, Italy, undertook a total of 130 inseminations, but only nine women had become pregnant. Worse yet, three of the female donors— who had been inseminated with the recipients' husbands' sperm—had remained pregnant after the flushing procedure, necessitating abortions. The low success rate and risks to donors might have been overlooked by FGR had the business not been hemorrhaging money. In early 1987, it showed a previous quarter loss of $316,289, which led Sucsy to announce in March 1987 that FGR would no longer be attempting to establish embryo transfer centers.

Nearly three years passed as we on the committee hammered out ethical guidelines. I continued giving speeches to infertility doctors, telling them about their potential liability for misleading women about their success rates. At one meeting, I advised doctors to stop using fresh semen for artificial insemination. The risk of transmitting the newly discovered AIDS virus was just too high. The virus could be transmitted in the sperm months before it could be detected in donor's blood, so there was no way to check if fresh sperm was safe. The proper approach would be to take the sperm, quarantine it in a freezer for six months, test the donor at that later time, and, if there was no virus in the donor's blood, use the sperm that had been collected earlier.

The doctors in the audience of that speech were furious that I was telling them how to run their practice. "I didn't screen my wife before I had a child with her; why should I screen donors?" one asked me.

Another area of contention in the Ethics Committee was whether it was "ethical" for postmenopausal women to use hormonal treatment and embryos donated from younger women to have babies. At one meeting, Howard Jones fought against the idea of women in their late forties and early fifties having children since the mothers might die before the kids reached college age.

I pulled an article out of my briefcase. Just that morning, the local paper had reported the happy news that Howard Jones had created an IVF child for a couple in which the husband was in his late seventies. Yet he was denying women in their late forties the right to have children because they might die young.

"My husband smokes a lot," I said. "What if he dies and I marry a younger man and want IVF? He'd be around to raise the child."

"Lori," drawled Richard Marrs, a young IVF doctor on the committee whose Texas accent made my name sound like it had a few extra syllables. "You couldn't get IVF. You'd never pass our psychological screening."

• • •

After spending nearly a decade of my legal career helping infertile women, I expected it would take me a long time to conceive, and I anticipated having to put myself in the hands of doctors such as these. I knew the dozens of ways that egg and sperm could go awry in their attempts to meet, merge, and continue their relationship as an embryo. The reproductive process was a stunningly complex one. "As a spectacle, it can be compared only with an eclipse of the sun, or the eruption of a volcano," wrote Dr. G. W. Corner in 1942. "If this were a rare event, or if it occurred only in some distant land, our museums and universities would doubtless organize expeditions to witness it, and the newspapers would record its outcome with enthusiasm."

I was shocked when, like some wayward teenager, I got pregnant almost instantly. The pregnancy test glowed blue one morning, right before I flew to London to join up with Patrick Steptoe, the obstetrician who was Robert Edwards's partner, to advise a meeting of international lawyers about the medical and legal vagaries of in vitro fertilization, donor insemination, and surrogate parenthood.

Happily in England, I wandered quaint streets near my hotel on Half Moon Street, thinking about the bundle of cells growing inside me. I was writing a book on surrogate motherhood and, in order to understand the bond between the baby and the surrogate, had been reading Thomas Verny's *The Secret Life of the Unborn Child*. Verney alleged that a fetus's experiences in the womb continued to influence him or her later in life. So now I had to worry not only about all the ways I could mess up a kid after birth, à la Freud, but also about how I might be depriving the baby culturally by not attaching earphones around my belly (as some mothers-to-be did) and playing classical music before birth.

I loved walking those British streets thinking about the child growing within me. I imagined that baby enjoyed visiting the same whimsical shops that I did. Later in my pregnancy, I learned the baby was a boy. I thought about the study by City University of New York sociologist Barbara Katz Rothman. She discovered that when couples learned the sex of their baby before birth, stereotyping began in the womb. If the baby were a boy, the mother described its movements differently than if it were a girl. A boy's movements are described as jerky and violent and a girl's as soft and rolling. Shower gifts for boys were all baby blue, welcoming the child into a monochromatic world— rather than the rainbow facing a child whose sex was unknown until birth. What's more, Rothman found that when the sex of the baby wasn't known until birth, mothers of boys were happier than mothers of girls. When the sex was known during pregnancy, the reverse was true. Women who were carrying a male were less happy, Rothman speculated, because they felt that there was something foreign within them.

I tried not to pigeonhole the baby or his movements before he was born. But the prenatal prediction had changed my view of the fetus in a way that was hard to grasp. Now that I knew the baby was a "he," I wondered, irrationally, if he really wanted to go to the same places I did during the pregnancy. Maybe whimsical shops were not to his taste.

I continued my practice and lecturing throughout my pregnancy. Airlines get a little panicky flying a pregnant woman— probably out of fear that she might go into labor, forcing an emergency landing—so I disguised my pregnancy with boxier and boxier clothes.

Once the truth was unavoidable, though, I learned how intensely people believe that a private pregnancy is a *public* matter. Strangers would reach out to touch my belly. People at restaurants would scold me when they thought the apple juice I was drinking was Chardonnay. Since I had testified in Congress *against* laws making surrogate motherhood a crime, my pregnancy attracted snide remarks from surrogacy foes.

Harold Cassidy, the attorney for surrogate mother Mary Beth Whitehead, who made headlines by changing her mind and wanting to keep the baby, approached me at Princeton University, where we were both speaking.

"Sell me your baby," he said.

He couldn't understand my view that surrogacy was right for some women in some of their pregnancies. It was as if he would assume that any woman who was pro-choice would abort every one of her pregnancies.

The infertility doctors were even worse when they learned I was pregnant. Annoyed at my telling them what to do for years, they now had an outlet for their resentment. When I attended meetings, they would gleefully tell me how much pain and suffering pregnancy would cause me. "Soon you won't be able to come to these meetings," an infertility doctor said to me, "because the fetus will be pressing so hard against your bladder all you will want to do is go to the bathroom every five minutes." Another told me that the fetus would probably press against my lungs so that I'd be gasping for breath. Pleased to be finally in a position to have specialized knowledge to dominate me, they tried to make me feel as if my body had been taken over by a satanic invader.

• • •

The month Christopher was born, a woman I had never met was running into a problem with Howard Jones's clinic in Virginia. Risa York had been trying to conceive a child for the previous four years. Because Risa had only one fallopian tube and it was damaged, she and her husband, Steven, had enrolled in Howard Jones's clinic and undertaken in vitro fertilization, in which her eggs were fertilized in a petri dish with Steven's sperm. On their fourth unsuccessful attempt at in vitro fertilization, they were offered the opportunity to have one of their embryos frozen for later use.

Before freezing the embryo, the Yorks were asked to sign a consent form. It said that the Yorks had "the principal responsibility to decide the disposition" of their embryo.

During the course of their treatment, the Yorks moved to Los Angeles, and decided to change clinics and obtain in vitro fertilization at the Institute for Reproductive Research there. They requested that their new infertility specialist, my fellow committee member Dr. Richard Marrs, implant their frozen embryo at his clinic. Marrs was well known and well respected, and in fact had achieved the United States' first birth from a frozen embryo three years earlier.

Steven and Risa called Jones's clinic to determine how to safely move the embryo from the East Coast to the West Coast. They learned that it could be transported—like other human tissue or organs—in a liquid nitrogen tank known as a biological dry shipper. Since Steve was a physician who had transported human tissue, such as corneas, while a medical resident, he felt completely comfortable about picking up the embryo and flying it back in the biological dry shipper in the seat alongside him on the airplane.

On May 28, 1988, Risa called the Virginia clinic and said that she wanted to make arrangements to pick up her embryo. The physician who took the call told her he would not allow it. Risa and Steve were stunned. "They're holding my baby hostage," Risa told her mother.

Her husband wondered about the extent to which economics played a role in the Virginia clinic's decision. When Howard Jones opened his clinic, it was the only game in town. Couples from around the country came to Norfolk, leaving behind frozen embryos. The clinic stood to gain thousands, maybe millions, of dollars if those couples had to return to the Jones Institute for reimplantation. That money would be lost if couples were able to move their embryos to the hundreds of other clinics that had sprung up.

Calls and letters from their new physician to their old one attempting to get permission to move the embryo were to no avail. In November 1988, the Yorks called me.

I normally did not get involved in private litigation, but their situation touched an emotional chord. In May 1988, when Risa had asked for her embryo, I had just given birth. I was experiencing all the excitement of new motherhood while she, it seemed, was being denied it. I agreed to work on the case without charging the Yorks a fee.

The couple's right to their embryo seemed unassailable. They had a contract right based on the language that they had principal decision-making authority over their embryo; a property right (a recent California case had held that a person has a property right to his or her genetic material, even if it is outside the body); and a constitutional right to privacy to make reproductive decisions, including the decision to transfer their embryo. I drafted a legal complaint, then contacted Jeremiah Denton, an attorney who practiced in Norfolk, where the Jones clinic was located. I needed a local lawyer to help me make the complaint more persuasive to the Virginia federal court. Denton made a suggestion, "Why don't we give the embryo a name?"

I blanched. If we humanized the embryo too much, the court

might declare the embryo to be a person. Then the court might want to appoint a guardian, and the defendants might be able to assert a claim to the embryo on the grounds that they knew better than the parents what was in the embryo's best interest. Turning the embryo into a person with legal rights of its own would set a dangerous precedent. Women's abortion rights would be in jeopardy; abortion would be deemed murder.

Jerry responded with a fax: "The advantage of giving the embryo a touch of personality is simply that it lends a lot of human interest to the story being told here. The first maxim of trial lawyering is that 'plaintiffs lose boring cases.' If this case is only about frozen cells belonging to a woman who could have lots of other frozen cells on ice somewhere, it doesn't have much sex appeal, either for the court or the wider audience we are trying to influence. If, however, it is about two cells that could someday become a person (in a legal sense and otherwise) and make the Yorks happy parents, being prevented from doing that by a powerful clinic whose sole interest is protecting its image, then we have the makings of a fabulous David and Goliath story.

"True, giving the embryo a name tends to humanize it, but the great philosophical and legal dispute is not whether the embryo is human, but whether it is a person in the constitutional sense, and I think the law is settled that it is not. Even a boat, a corporation or a dog can have a name. Why can't Embryo Y?"

But I still wasn't convinced. Whether an embryo was declared a person might depend on something as uncontrollable as a particular judge's personal or religious background. I forced Jerry to delete the name. Two days after Mother's Day 1989, we filed a brief referring to "the Yorks' embryo" or "their embryo," emphasizing the Yorks' claim to the embryo rather than the embryo's own rights.

The day before the hearing, I flew to Virginia and met Denton to draft a final version of the reply brief and prepare for the next day's oral argument. The defendants had made a claim that allowing the Yorks to get their embryo back would open the door to Brave New World abuses. It seemed to me, though, that if anyone were guilty of a Brave New World abuse, it was the physicians. After all, in Huxley's book, part of the problem was that the directors of the clinics took control from parents over the fate of test-tube babies. We asked the hospital in which Jones's in vitro fertilization lab was located to aid us in getting the Yorks' embryo back, but the hospital's lawyers claimed it would be impossible. They said the clinic protected the embryos in a high-security tank and the hospital did not have a key.

In the months during which the judge was considering the Yorks' request, the American Fertility Society Ethics Committee continued meeting. Howard Jones, Richard Marrs, and I sat in small conference rooms at fancy hotels with the handful of other members of the committee and tried to develop rules for the profession. Jones, as chairman of the committee, put on the table the issue of couples transferring embryos to other clinics. He was planning to do an end run around the court by having the AFS declare it unethical for clinics to transfer embryos to other clinics.

Eventually, the judge ruled that the Yorks had valid causes of action under contract, property, and constitutional law. Three months later, the clinic finally agreed to give the embryo back.

The next morning, television cameras tracking our every step, Steve and Risa York, Jeremiah Denton, and I went to the clinic to rescue the embryo. The clinic would only allow one of us—Steve—into the room in which the embryos were stored to witness the transfer of the Yorks' embryo from its home in holder 2, cane 3 of storage tank 1 to the traveling biological dry shipper.

"Do you remember how the hospital described the super security measures preventing them from helping us gain access to the tank with our embryo?" Steve asked when he reappeared. "Well, all the tank has is a tiny lock, like the kind you get on cheap luggage, the kind you can open with a hairpin."

We proceeded to the airport, where the Yorks had reserved three seats across for their journey back to California. They put the three-foot cylindrical dry shipper into the window seat, put a pillow around it, and clicked its seat belt.

• • •

Because of enduring pro-life sentiment, there are still no federal funds available for procedures involving embryos. Only a dozen states have laws requiring insurers to pay for in vitro fertilization or related procedures. This means that clinics are in a fierce competition for wealthy patients. Some clinics report as "pregnancies" small hormonal shifts in a woman's body, which show that an embryo had briefly implanted—before being reabsorbed by her body. Others implant as many as ten embryos or use infertility drugs indiscriminately to increase the number of babies the clinic created, even though this increases the risk to the woman and the fetuses.

Infertility services have been transformed from a small medical speciality to a $2-billion-a-year industry. Couples seeking IVF are spending $44,000 to $200,000 to achieve a single pregnancy. Infertility specialists are now the highest-paid doctors, with experienced ones making an average of $625,000 per year. Even IVF pioneer Robert Edwards has gone commercial. In

1986, he sold Bourn Hall to Serono, the drug company that makes fertility drugs, for a rumored $12.6 million. "The company now has a window on reproductive technologies in a way we never had before," Serono's Thomas Wiggans told reporter Robert Lee Hotz. "We always had good relationships with clinicians, but, to be blunt, we didn't own them. Now we do."

The array of reproductive options has grown: egg donation, transfer of embryos from one woman to another, grandmothers gestating their in vitro grandchildren when their daughters can't carry the pregnancies. More than 600,000 Americans have tried reproductive technologies.

But battling sterility is not the only interest of scientists experimenting with conception. By moving reproduction into the labs, scientists face an unprecedented opportunity to shape the way human beings develop. They can make "improvements" on nature by choosing eggs or sperm from donors with desirable traits or, potentially, can even rewrite a child's genetic code to develop traits that may be completely new.

Many clinics are using novel fertility procedures in women without telling them how experimental they are. One infertility doctor boasted to me: "We go from mindside to bedside in two weeks. We make things up. We try them on patients. And we never get informed consent because they just want us to make them pregnant."

The right to be surreptitiously experimented on certainly wasn't what I was fighting for when I started representing women like Risa York.

3

Multiple Choice

❧

In November 1997, Bobbi McCaughey gave birth to septuplets. The newspapers called it a "miracle," but that hardly seemed an appropriate term for a doctor-induced superpregnancy that could have led to eight deaths, endangering both Bobbi and the babies.

"Women's bodies aren't designed to give birth to a litter like an animal," I told Ted Koppel on *Nightline*.

My mother phoned, chastising me. "That wasn't a nice thing to say," she pointed out. "And they just look so *cute*."

Her opinion of the incident changed, though, when I explained that it wasn't God or even fertility drugs that left Bobbi facing a future of 35,000 diapers, but something more akin to medical malpractice. In most cases, supertwins—as multiple births of triplets and beyond are called—are avoidable.

"Here we had a nice ending," said Mark Sauer, medical professor and chief of reproductive endocrinology at Columbia University in New York. "However, if she would have died or the babies had lifelong problems with cerebral palsy or some other complications, it would be a nightmare. Even now, with the potential for psychosocial problems, it may still become a nightmare. And society certainly has the right to ask, 'Why did this happen and were there alternatives?' "

"When people see the septuplets and talk about medical miracles, that is not the way I see it," says Dr. Daniel Kenigsberg, codirector of Long Island IVF. "I see it as a complication. Sometimes we do too good a job."

Fertility drugs are only a generation old. Thirty years ago, human menopausal gonadotropin, a hormone, used to create the drug Pergonal, was collected in Italy from the urine of post-menopausal women. A prime source was older nuns, who, with religious vows preventing them from having their own children, readily agreed to facilitate someone else's pregnancy.

With a good doctor, Pergonal is injected into the woman to cause the ovaries to produce multiple eggs. By doing hormonal blood tests and ultrasound exams to learn how many eggs are likely to be released, the doctor can decide whether or not to give the subsequent injection of another hormone to release the eggs. If there are too many, the doctor will tell the couple not to attempt pregnancy that month—the risk of multiples is just too high.

At first fertility drugs were only prescribed for women whose ovaries did not function. Today they are prescribed to women who are merely impatient—those who haven't gotten pregnant after a few months of trying, and whose ovaries virtually burst forth eggs when the drugs are administered. In 1997, nearly 1.3 million fertility drug prescriptions were filled, at a cost of $230 million.

So many women are given infertility drugs that, in 1995, a shortage of the $2,000-a-month Pergonal was declared in the United States. Doctors argued the shortage was creating major problems for women in their late thirties and forties who were using the drug and were trying to beat their biological clocks. The U.S. Food and Drug Administration took the unusual step of allowing foreign import.

But fertility drugs are dangerous medicines, presenting grave risks to women and children. Yet there is nothing that legally limits their use to certain types of patients; nor are they limited to infertility specialists. In fact, any family physician can use them. I recall sitting in an L.A. diner with an infertility specialist as he was beeped by a GP calling from another state who had given a patient a cocktail of fertility drugs, and now saw that her blood pressure was falling and wanted instructions on what to do.

While 58 percent of multiple births are delivered by women who use fertility drugs, another 22 percent are born of women who undergo IVF. In fact, one in three IVF births produce multiple children. Doctors generally implant more than one embryo in the woman, expecting that not all will develop into a pregnancy. In California, a forty-nine-year-old woman who had seven children and six grandchildren in her first marriage decided to undergo IVF with husband number two. Five embryos were implanted, which doctors hoped would lead to one child. Four babies were born.

Managed care plans—and certain insurance policies—only cover a certain number of IVF attempts. So the incentive is to try to have as many babies as possible in those limited attempts. "If the couples have failed at an IVF attempt, they have a gambler's mentality where they ask for even more embryos the next time," says IVF doctor Sauer.

Clinics use fertility drugs and multiple IVF embryos to inflate their pregnancy rates. If a clinic has 200 patients, it can report the creation of 150 babies, without mentioning that 50 patients had three babies each, and the majority of patients went home without a child. And in this competitive age of *New York Times Magazine* ads by fertility clinics, some doctors jump the gun on fertility drugs to get more patients. Some doctors give fertility drugs to women who have been trying to have a child for just three months, even though perfectly healthy fertile women often take up to a year to become pregnant.

Until every major media outlet called me for a comment on the septuplets, I hadn't given much thought to the issue of multiple births. Once I started looking into the subject, though, I wondered why the Centers for Disease Control hadn't declared multiple births a major public health problem. Fertility doctors were taking unnecessary risks with the bodies of women and children, and the results were straining hospitals and neonatal intensive care units to the breaking point.

For the first fifty years of the century, there were forty-six sets of quadruplets in the world. Now that many are born each year in the United States. In 1996, there were 100,750 children born in the United States as twins, 5,298 born as triplets, and 641 children born as higher-order multiples.

When asked about his run of multiple births, Dr. Donald Young of the Mid-Iowa Fertility Clinic said, "Certainly they're a success for the couple."

When women tell Dr. Victor Klein at North Shore Hospital on Long Island that they are troubled by having triplets or quadruplets, he says, "What do you mean you're not happy? This is everything you wanted, and then some."

Yet overstimulation of the ovaries by fertility drugs can cause swelling and bleeding of the ovaries and severe fluid retention—and, in some cases, heart failure. In addition, a woman pregnant with more than two fetuses is at risk of potentially fatal blood clots and diabetes. She may require bed rest, hospitalization, medications to stop early contractions, or cerclage, a procedure where the cervix is sewn shut. Women on fertility drugs are also at an increased risk for ovarian cancer. The FDA now requires that all fertility drug labels disclose the cancer risk.

Many clinics use consent forms that list totally remote possibilities—what would happen to an embryo if there was an earthquake, act of God, labor strike, or war—but not the very real (and statistically much more probable) risk of multiples. Some clinics never mention that one in three IVF births is a multiple.

Imagine the impact on the fetuses themselves from sharing those tight quarters.

During development in the womb, fetuses in a multiple pregnancy are at a higher risk for lung development problems and cranial hemorrhaging. When multiple embryos are put in the womb, there is a danger the embryos may fuse, creating what is known as a chimera. In one case, a male IVF embryo and a female IVF embryo put into the mother's womb at the same time fused into a single individual, who had both male and female sex organs at birth. Doctors suggest that other chimeras may have occurred but gone unnoticed because the embryos were of the same sex.

While only 8 percent of singletons are premature, the percentage rises to 53 percent for twins and 92 percent for triplets. Seven percent of babies in multiple births die within the first year after birth, compared to 0.6 percent of singletons.

If they do survive, multiple-birth children may have longer hospital stays than singletons. Only 15 percent of singletons require neonatal intensive care, compared to 78 percent of multiples. The McCaughey septuplets didn't all make it home until three months after their birth. The youngest, Alexis, came

home with an oxygen tank. She later returned to the hospital
with breathing problems, and her brother Kenny had eye
surgery. Multiples may also suffer long-term medical problems,
including lung disorders, cerebral palsy, blindness, and learning
disabilities. A set of sextuplets in Indiana required three years of
state-funded special care. A set of sextuplets in New York also
has medical problems. One baby is legally blind, another has
one bad eye, and another has epilepsy.

Yet no registry tracks the children to measure any problems.
The full magnitude of the risks is unknown.

Even if many of the dangers associated with multiple births
could be reduced, the health care costs associated with supertwins are high. In 1991 the cost of labor and delivery for a single
child was about $10,000, while the cost of having twins was
$38,000 and that for triplets exceeded $100,000.

Hospitals are burdened by the multiple births and may need
to expand their facilities to include larger neonatal intensive
care units (NICUs). Edward Hospital in Naperville, Illinois,
for example, is expanding its NICU because of the recent
increase in multiple births in this high-income town of thirty-
somethings using fertility drugs. My local drugstore, in a
trendy area of Chicago, has a Web site about fertility issues
because of its desire to attract the wealthy women who can
afford these prescriptions.

The McCaughey births required forty health care professionals, divided into seven teams, at an estimated cost of up to
$1 million. The delivery was like a military exercise, with teams
A–G going through their drills moments before the birth.
Intensive care nurse Cecelia Kirvin told reporters: "I would call
out 'Baby A, sound off,' and the team for Baby A would tell me
if they had any equipment needs or last minute questions. Then

I did that for Baby B and all the other teams." When it was over, all four boys and three girls survived: Kenneth, Nathaniel, Brandon, Joel, Kelsey, Natalie, and Alexis.

At the moment, it is the neonatal intensive care nurseries that have to cope, but in a few years the phalanx of supertwins will enter the schools. In small school districts, multiple siblings cannot be divided into different classes to encourage them to interact with other children and develop socially.

Multiples start talking later, maybe because they do not have as much individual attention, or because they develop nontraditional ways of communicating with each other. Sibling rivalry may be stronger because the children have to compete for their parents' attention. Multiples are more likely to be compared to each other than siblings born years apart.

The parents of multiples are more apt to be exhausted, depressed, anxious, and isolated from other couples and old friends. Imagine trying to teach five or six children to use the toilet at once, or read at once, or years later, to drive at once. As many as one-third of the parents of multiples split up before the children reach the age of three. Having multiples is also a tremendous financial strain that only grows as the years pass.

Keith and Becki Dilley, the parents of sextuplets, took extra jobs but still had to sell their house and move to a smaller one. Keith Dilley said, "There are days where you just can't wait for that day to get over. You put them to bed that night and you just go, I hope tomorrow's better."

One way to minimize multiple births is to pass a law limiting to three the number of IVF embryos that can be implanted in the woman at one time, as has been done in England. In the United States, the American Society of Reproductive Medicine (as the American Fertility Society is now called) recommends such a limit, but compliance is voluntary. I know doctors who put in seven or ten embryos.

For the woman who produces only two or three embryos, infertility doctors are looking into the possibility of splitting those embryos in half or quarters to enhance the chance of a pregnancy—a technique learned from the cattle industry— perhaps further increasing the incidence of supertwins. Breeders learned that a superior cow embryo worth $1,500 could be divided into eight to sixteen parts, each creating a separate, commercially valuable new embryo to be put into less valuable cows. In humans, the embryo generally would be implanted in the original mother.

Doctors say they implant numerous embryos to increase the woman's odds of becoming pregnant. But a study reported in August 1998 in the *New England Journal of Medicine* showed that implantation of two embryos is optimal—the chances of pregnancy do not increase beyond that.

Like the proverbial traveling salesman who impregnates the farmer's daughter, the infertility doctors who use technology to create superpregnancies seem untroubled by their results. They offer another technology: the termination of some of the fetuses. The physician injects potassium chloride into the heart of one or more of them. Usually the other fetuses survive, although the chance exists that the entire pregnancy will be miscarried. Doctors avoid the term *abortion*. They call the procedure *selective reduction*.

Bobbi McCaughey was, of course, offered the option of "selectively reducing" some of her seven offspring. Doctors were shocked when she refused because of her deeply felt religious beliefs. Yet it is perfectly understandable that an infertile woman could find it emotionally wrenching to terminate lives she and her husband worked so hard to create.

Infertility doctors don't pay sufficient attention to the risks of multiples because they aren't confronted with the problems they've created. The in vitro fertilization doctors don't do the selective reductions themselves. The infertility doctors are not in the neonatal intensive care unit when these children are born suffering from blindness, neurological defects, or cerebral palsy.

Should high-tech abortion be available to compensate for the problems caused by high-tech reproduction? Denying that right, said Dr. Mark Evans of Detroit, who performs one hundred selective reductions a year, would be like denying medical attention to crash survivors because they voluntarily got into the car. "We as a society are trying to make the lives better," he says, "for the people in it . . . and that involves the use of technology."

Selective reduction itself has risks. A year after the procedure, one-third of women still have persistent depressive symptoms related to the reduction, mainly sadness and guilt. In the first five years it was used, from 1984 to 1989, one-third of the women undergoing the procedure lost *all* their fetuses. Even now the pregnancy loss rate is between 7 and 13 percent. And many of the fetuses who do survive are born prematurely.

"You do damage whenever you go in with a needle and leave dead tissue behind," says University of Michigan epidemiologist Barbara Luke. "The long-term effects on these other children, who knows?"

• • •

Supertwins can be like a traveling sideshow. "What are those?" people ask the Dilleys about their sextuplets.

"Children," Mr. Dilley explains.

Since the 1934 birth of the Dionne quintuplets, supertwins have been turning heads. Annette, Cecile, Emilie, Marie, and Yvonne Dionne—identical quintuplets from the same embryo—were the subject of ruthless exploitation. The Ontario, Canada, government got a court order ousting the Dionnes' own parents and making the government their guardian. Then it turned the quints' home into a theme park. Quintland was visited by millions, including Clark Gable and Amelia Earhart. By the time they were nine, the girls had been allowed outside the compound only three times. Their doctor was the best-known physician in the world.

Dionne quint dolls were more popular for a time than even Shirley Temple dolls, and their image was used to sell baby food, disinfectants, toilet tissue, and other products. Most of the money went to the government. The exclusive contract between the government and an official photographer prevented the quints' dad from taking pictures of them.

The quints were put under immense scrutiny—not only by being displayed three times a day to over 6,000 people, but by researchers who studied their every tantrum and bowel movement.

When the McCaughey septuplets were born, the three surviving Dionne quints wrote to them. The letter, published in the *Los Angeles Times*, stated: "We would like you to know we feel a natural affinity and tenderness for your children. . . . Multiple births should not be confused with entertainment, nor should they be an opportunity to sell products."

Until recently, the surviving quints—Annette, Cecile, and Yvonne—were living on a joint pension of $490 per month. In March 1998, sixty-three years after their birth, they received a $2.8-million settlement from the Ontario government to compensate them for their exploitation.

The Dionne quints' life looks not so much like a horror story as a career plan to certain couples using fertility drugs. When Kim and Lauren Forgie had quints with the help of fertility drugs in Canada a decade ago, they hired Price Waterhouse to weigh the commercial offers they thought would come. When they realized there was little commercial interest, they hired a public relations firm to generate some. The result was a one-year advertising contract with a children's clothing manufacturer.

In England, Mandy Allwood fell in love with Paul Hudson, a self-described womanizer. She left her husband and got pregnant with Paul's child, but miscarried. Four months later, she persuaded her GP to give her fertility drugs. As the mother of a child conceived with her husband, and someone who had recently been pregnant, she certainly didn't fit the profile of "infertile." She was simply in a rush.

Pregnancy came swiftly with the drugs, and it looked like there were multiple fetuses.

Paul hadn't completely committed to Mandy; he was still living half of each week with his old girlfriend. When he thought Mandy was pregnant with four babies, he wasn't amused. But then Mandy had further prenatal testing. When she told Paul she might have seven fetuses, he perked up. "Well, that's interesting," he said. He got even more enthusiastic when he learned there were eight.

Since eight babies had never survived the womb, the doctors offered her selective reduction. But a British tabloid made a better offer. Mandy and Paul would be paid for exclusive rights to the story—with the size of the payment commensurate with how many babies survived.

At nineteen weeks, Mandy could feel the babies kicking. She needed to make it only a few more weeks for them to survive. She appeared—for a fee—on the *Rolanda* talk show, but by the time it aired, she had begun to miscarry.

Over the next three days, she lost the babies, one by one. But she and Paul made enough money in the brief media frenzy to buy a home in London and live comfortably for two years.

The McCaugheys, too, have learned that multiples can mean money. The Iowa governor promised them a home; a 6,000-square-foot house with seven bedrooms and two laundry areas was ultimately built with private donations. Toyota gave them a minibus. They received seven free years of cable television, a sixteen-year supply of applesauce, and a lifetime supply of Pampers. And their hometown of Carlisle, Iowa (population 3,500) volunteered baby-sitting services.

They have an agent now—who handles Christian clients—to weigh offers for them. Kenny McCaughey, the dad, got the lead in a power tool commercial, and Bobbi, the mom, got a deal designing baby clothes for Simplicity Pattern Co. They sold the movie rights to their story. And, as with the Dionne quintuplets, the McCaugheys sold the rights to every photo of the babies for two years to a New York company.

It takes Bobbi three hours each day to open presents sent by well-wishers from around the world. But as I read about each new gift for the McCaugheys, I think about the thousands of couples who have triplets, quadruplets, or more, courtesy of promiscuous infertility doctors. No governor has offered to build them homes. The neighbors aren't volunteering to baby-sit. They can't even count on a disposable diaper manufacturer for a break.

The companies have become jaded. One company spokesman told a mother of quadruplets, "Everybody has quads. We don't do anything for multiples anymore. If you are going to have six, let us know."

4

Souls on Ice

❧

Loretta and Basilio Jorge had struggled with infertility for years. They went to the University of California at Irvine to consult Dr. Ricardo Asch, a world-famous specialist whose license plates read DR. GIFT, the acronym for gamete intra-fallopian transfer, the infertility treatment he had pioneered. A regally handsome Argentine whose hobbies include playing polo, Asch had succeeded in doing what no other infertility specialist had done—developing a procedure that even the Vatican blessed. Instead of fertilizing the woman's egg with her husband's sperm in a plastic dish, Asch injected them into the woman's fallopian tubes, where fertilization could occur more naturally.

An Asch colleague told me, "Dr. Asch is one of the greatest experts in infertility in the world. He's an innovative scientist, and a generous teacher." The husband of a patient said, "I would give him my life. This country has to make a statue for the guy."

But in time, whistle-blowers began revealing another side of the doctor. They told university officials about suspicious activities at Asch's clinic. They claimed he used a drug on patients that had not been approved by the FDA. They also made a stunning accusation: that Asch had stolen eggs from some of his infertility patients and, without their knowledge or consent, implanted them into other patients who were duped into thinking they were from legitimate donors.

The claims were shocking and serious. But according to a lawsuit later filed by a patient, rather than firing Asch, the university reportedly paid $900,000 to the whistle-blowers not to divulge.

The Jorges had been among Asch's first patients in Irvine, spending $30,000 on infertility treatment. Loretta became pregnant but miscarried after three months. Unbeknownst to Loretta, five of her eggs had been implanted in another woman. That other woman gave birth to twins.

Six years later, the news reached the still-childless Loretta that two children—a boy and a girl—had been born using her eggs. Her husband, Basilio, was desperate to see if their father was treating them well. Loretta wanted to visit them. The Jorges videotaped the children waiting at their bus stop. They played the four-minute video again and again. Loretta noticed the girl walked just as she did when she was a girl. Growing ever more fond of these six-year-olds, they decided to take legal action: they filed a lawsuit seeking custody of the twins.

The Jorges were not alone. Deborah and John Challender learned that they also had twins born to another woman. They were troubled that the twins were being raised in the Jewish faith, rather than following their beliefs as Christians.

Renee Ballou unwittingly donated an egg that resulted in a live birth, though she did not conceive. She is haunted by the fact that she may have been commiserating in Asch's waiting room with the very woman who now is raising her child. Ballou believes that her inability to conceive caused her divorce. Currently engaged to be remarried, she wants the child born to the other couple to escort her down the aisle.

"He was playing God," Ballou said angrily of Asch. "He basically took my hopes and dreams and gave them to someone else."

Over a hundred couples were affected, resulting in eighty-four lawsuits against Asch, his colleagues, and the university. The couples claimed that eggs of the wives or, in some instances, embryos of the couples, had been stolen and given to other patients, sometimes resulting in children who now lived in other countries. A Santa Ana couple learned four of their fertilized eggs had been sent to the University of Wisconsin for research by a zoologist.

In response to the lawsuits, Asch fled to Mexico.

Such theft was "predictable, almost inevitable," says Boston University health law professor George Annas. "The field of reproductive technology is so lucrative and so unregulated that someone was just bound to do it."

A lab assistant from Asch's clinic stated in her deposition that his decisions to transfer eggs and embryos without consent were "deliberate," but that she was unable to challenge him "because she feared retaliation." Dr. Asch defended himself, telling *Time* magazine, "I think there were people in charge of setting me up who falsified documents and forged consent forms." Former chief biologist Teri Ord corroborated the lab assistant's accusations, saying that Asch not only knew of the misappropriations, but ordered them.

Asch was eugenic in his selections as well, tending to take eggs from tall, blond-haired, blue-eyed women. He may have taken eggs from younger women and used them in older women to inflate his success rates.

Closing the barn door after the horse had fled, the University of California at Irvine invited a panel of experts, including me, to advise the administration on the proper guidelines for oversight of infertility clinics. The California legislature passed a law, California Penal Code Section 367, effective January 1, 1997, making it a crime to steal someone's eggs or embryos.

But why did the university need me or the legislature to tell them you shouldn't use a woman's eggs to make babies without her consent?

"Making someone a parent without their permission is akin to crimes like rape and assault," says Dr. Arthur Caplan, a University of Pennsylvania bioethicist.

The university agreed to settle the suits brought against it by former fertility clinic patients. Its lawyers developed a cold actuarial formula to determine how much a given couple had suffered. The most money was given to couples who didn't get pregnant but whose eggs or embryos resulted in children for other couples. The next category, worth less money, involved situations where both couples got pregnant. Even less money was given to women whose eggs or embryos were miscarried by the second woman or whose eggs or embryos were improperly used for research.

The Jorges clung to the videotape of the twins, showing it to twenty members of their family on Christmas. They offered to pay half the child support and requested joint custody with the birth parents.

In 1997, the university settled with the Jorges for $650,000, part of $18.4 million the school gave to sixty-one couples.

Amazingly, on February 2, 1997, after trying to have a child for more than sixteen years, Loretta Jorge gave birth to a baby boy. Joshua James Jorge, whose name, Loretta says, means "God is my savior," was conceived without fertility treatments.

The university had paid for its lack of oversight, and has since instituted more stringent protocols for its infertility clinics. But the main culprit accused of wrongdoing, Ricardo Asch, was not reached by the litigation. As of this writing, representatives from the California Medical Board, the Federal Bureau of Investigation, the U.S. Food and Drug Administration, U.S. Customs, the Internal Revenue Service, the Defense Criminal Investigative Service, the University of California at Irvine Police Department, and the National Institutes of Health are all trying to decide what to do about him.

Extradition from Mexico seemed like the first step. But again, the uncertain status of reproductive technologies came into play. You can only extradite someone for a felony, which, in this case, would be a theft of property over $50. But since our law and culture refuse to put a price tag on human embryos, the case might not qualify.

So Dr. Asch now oversees a flourishing infertility clinic in Mexico. He also was named international editor for Robert Edwards's scientific journal *Human Reproduction*.

• • •

"I have three hundred unclaimed embryos in my freezer," one South Florida IVF doctor told me. "I've tried tracking down the couples, but they must have moved. Can I give them to other couples in a prenatal adoption? Or should I just terminate them?"

I tried to imagine the public relations impact of trashing three hundred human embryos. The newspaper headline "Massacre Near Disney World" came to mind.

On one hand, human embryos are the most valuable and revered entity on earth. An entire social movement—the right-to-life movement—is dedicated to protesting, harassing, and even killing in order to protect their right to be born. Infertile couples are willing to spend $16,000 or more per attempt to create an embryo. Researchers see embryos as a Xanadu of research possibilities, as a source of cells that can be used to treat Parkinson's disease or Alzheimer's.

And yet, in IVF clinics across the country, there are stockpiles of abandoned embryos. How can couples "forget" they have a frozen embryo? Isn't that like a man conveniently forgetting he has a wife?

Zoe Leyland, the first child born after being frozen as an embryo, entered the world on March 28, 1984. Zoe, a five-and-a-half-pound girl, had spent two months suspended as an embryo in an Australian clinic. By the following year, there were 289 embryos frozen in the United States. There are now more than 100,000 such souls on ice in the United States, increasing at a rate of nearly 19,000 per year.

The fate of frozen embryos is a major source of social debate. Couples freeze embryos because the fertility drugs given to women create too many embryos to safely implant. If the initial IVF attempt doesn't work, the couple can defrost and implant the extra embryos more cheaply—and less painfully—than if they began the whole cycle (fertility drugs, egg retrieval, etc.) all over again.

The fertility drugs given to Kathleen Markland allowed doctors to harvest enough eggs to create eight embryos. She had four implanted, giving birth to triplets. Now her clinic is pressuring her to decide what to do with the four remaining embryos. With three three-year-olds at home, she feels too emotionally exhausted and financially stressed to expand her family. But after spending six desperate years trying to have a child, she doesn't feel right telling the clinic to pull the plug on her frozen embryos.

Markland was not comfortable with outright termination, nor with donation of the embryos to other couples. So she came up with an idea that she felt would work best: her doctor could implant the embryos in her, but without all the hormonal and other treatments to maximize their chance of taking. If she got pregnant, fine; she'd love the baby like her others. But she would be making it far less likely a pregnancy would occur.

Markland asked her doctor if he could implant the embryos in her in a nonoptimal way.

"No," he purportedly told her, "that will ruin our statistics."

Clinics differ in how they handle the disposition of frozen embryos. One clinic only allows embryos to be frozen for two years. That doesn't make sense, though. If the woman gets pregnant with fresh embryos in her initial IVF attempt and gives birth nine months later, she may not be ready for a second child when her first one is just one year and three months old.

In another clinic, one doctor is opposed to the termination of embryos. He has exerted enough pressure to be able to block the other doctors in the hospital who are willing to provide the service. A couple who thought their family was complete and wanted to terminate their remaining embryos was told that the clinic would give the embryos to them in an envelope to take home and terminate themselves.

Such callous disregard for patients' feelings makes one think,

There ought to be a law. But in England, a law was adopted to govern the fate of human embryos—and its application was equally troubling.

The British Human Fertilisation and Embryology Act said that frozen embryos should not be stored for longer than five years. Physicians who failed to destroy the embryos on schedule faced jail, fines, or both. The original expiration date for embryos already stored was July 31, 1996. The government amended this provision in May 1996 to extend the maximum storage time as long as both "parents" consent to the longer storage.

The Human Fertilisation and Embryology Authority began reminding clinics of the impending disposal deadline in 1995 by sending letters to sixty-one clinics requesting information from each one about how many embryos it had in storage. It also warned the clinics that "the deadline cannot be ignored." The Department of Health began working with the HFEA to gather information about the number of embryos that were unclaimed.

Some of the embryos in storage had been frozen for over ten years. Clinics in Britain annually send patients a letter asking if they want to continue to store the embryos, donate them for medical research or to another couple, or have them destroyed. In 1995, clinics reported that up to 10 percent of the people contacted did not answer the letters or pay for the storage, but the clinics continued storing the embryos.

The clinics were unable to contact the patients because many of them had vanished or divorced and did not leave directives as to what should be done with the embryos. As the spokeswoman for a pressure group named Comment on Reproductive Ethics (CORE) put it, "The idea of all these babies around without anyone knowing who they belong to is quite unbelievable."

In some cases, though, couples who had fertility treatment in the 1980s before the law was passed might not have known about the law to destroy the embryos. One American couple was transferred from England back to the States by the Air Force. But the clinic couldn't approach the Air Force for a new address since it would have breached the couple's confidentiality to have identified them.

The doctors pointed out that patients return for many reasons to reclaim embryos. One woman recontacted the clinic after her child was killed in a car accident. Younger women especially might want to store their embryos for a longer amount of time. The clinics were also concerned about potential legal problems that could arise from destroying embryos without the informed written consent of the patients.

The clinics had difficulty contacting patients who lived abroad. They also had problems locating couples because the clinics were prevented from compromising the privacy of the couples by giving their names to private investigators or former employers who might have a forwarding address. One clinic used a CD-ROM database to search through the 44 million names on the UK electoral register but found only 29 of the 150 couples that the clinic still needed to contact. Frustration remained high. As one doctor said, "Are the couples going to say, 'Sorry, I forgot to call'? That is irresponsible."

The law did not make it clear who owned the embryos or what would happen if a couple was estranged and did not agree about the use or destruction of the embryos. Another glitch in the law penalized women with infertile husbands who had used sperm donors to create the embryos. Some such couples wanted to have more children with embryos from the same donor (so all their children would have the same genetic parents), but they didn't want them just yet. However, the law required that both genetic parents give permission for the embryos to be stored beyond the five-year deadline. In the case of eggs that were fertilized by donor sperm, the anonymous genetic fathers could not be contacted because it would violate privacy laws. The women had to use the embryos immediately or let them be destroyed.

Many of the clinics did not want to destroy the embryos but also were not able to indefinitely allocate the money and storage space to continue holding the embryos. Peter Brinsden, the medical director of Bourn Hall Clinic, who had taken over for Robert Edwards, said he would rather go to jail than destroy a single embryo without the couple's permission. "Each embryo has the potential of becoming a child even if it's a 10 to 20 percent chance of a baby arising out of that embryo," he said. But the HFEA told him the penalty would not be jail. Instead, it would revoke the licenses of Bourn Hall and all its doctors, leaving thousands of infertility patients without medical care.

The Vatican condemned the destruction of the frozen embryos, calling it "prenatal massacre." A British right-to-life group asked, "What is respectful and sensitive about throwing these tiny human beings into incinerators along with dirty swabs and bits and pieces from operating theatres?" The Vatican suggested that married women volunteer to bring the embryos to term in "prenatal adoptions" that would be similar to taking in an orphan or abandoned child. More than one hundred Italian women volunteered to adopt embryos, including two elderly nuns.

In spite of all the controversy, clinics in Britain followed the law and destroyed any abandoned embryos that had been in storage for more than five years as of August 1, 1996. At Bourn

Hall, nearly nine hundred embryos were scheduled for termination, a process that took two full days.

A chain of Italian clinics had offered to take the Bourn Hall embryos, but director Peter Brinsden declined. "How would people react," he asked, "if they found out two or three of their children were running around Italy in a few years' time?"

Prime Minister John Major denied a last-minute appeal by antiabortionist groups to delay the disposal of the embryos. He said that the destruction was the "will of Parliament." In Italy, the speaker of the Italian Parliament turned down a request from Alessandra Mussolini—the dictator's granddaughter—for a moment of silence to mourn the destroyed embryos.

More than 3,300 embryos across England were removed from storage. Some were allowed simply to defrost, while others were placed in a solution of vinegar and alcohol and then incinerated. Patients called up clinics until just hours before the deadline passed and asked that their embryos be spared. Clinics also received requests after the embryos had been destroyed. The doctors were understandably distressed. "I never imagined I would be destroying embryos where there are clearly women who are desperate for us to preserve them," said one.

• • •

In the fall of 2001, the Ariane 5 Space Rocket is scheduled to launch into outer space from French Guinea. For $500, you can ensure that a strand of your hair is on it. The sponsoring group, Space Systems at Encounter 2001, expects to launch hair from 4.5 million people, broadcasting digitally into space the names of the donors whose biological material is on the way.

The group already launched the ashes of LSD guru Timothy

Leary and *Star Trek* creator Gene Roddenberry. But some scientists are asking: Why stop at ashes and hair? Why not send human embryos as well?

The idea would be to colonize a planet by sending a few adults, some artificial wombs, and a selection of human embryos. The stumbling block has been the artificial wombs. But, in 1997, Japanese researchers announced that they were close to developing "ectogenesis," an artificial womb with the potential to bypass the need for human gestation.

The artificial womb lacks certain of the elegant design features of its human female counterpart. It is a clear plastic box of warm amniotic fluid in which a fetus is attached to a dialysis machine that replaces oxygen and cleans the fetus's blood. So far, the Japanese researchers have only tried the device in goats, removing the fetuses from their mothers three weeks before their due date and placing them in the artificial wombs. These fetal goats continued their gestation in the artificial womb until they were removed, or "born."

A Japanese biotech company plans to go even farther. In 1998, it began a study to grow cows outside the womb. Its researchers plan to take cells from the wombs and placentas of cows and grow them into a genetically enhanced placenta. At first, they will implant the new placentas into cows to lower the risk of miscarriage, but ultimately they will create an independent gestational device.

Researchers are already suggesting that a human embryo—created through in vitro fertilization or washed from a woman's uterus—could be implanted into an artificial womb and develop until "birth."

The artificial womb could be a great equalizer, making women and men indistinguishable in the ability to beget and bear children. Society could no longer presume women were responsible for childbearing because they could always "hire" an artificial womb to circumvent pregnancy. A man could buy embryos and use the machine, allowing himself to be a truly "single" parent.

But the artificial womb could also be used *against* the fetus's progenitors, depriving them of their right to terminate embryos. In the United States, there is no way a law based on the British model requiring termination of embryos would pass. Quite to the contrary, laws forbidding embryo termination and perhaps, in the future, requiring artificial gestation are much more likely.

Already, a Louisiana law prohibits termination of in vitro embryos. The law declares such embryos to be "juridical persons" and requires that each embryo be registered through something akin to a mini–birth certificate. If a couple believes that they have enough children and do not want to use the embryos themselves, apparently the state can swoop in and give the embryos to other couples.

But how would the genetic parents feel about a child of theirs being raised by someone else? Would they worry, as the Jorges did, about whether their baby was being properly cared for? Would they, like the Challenders, be troubled that the child was being reared in a different faith? And what if their own IVF child died—would they sue the recipient couple for visitation rights to the one created with their transferred embryos?

I advised Louisiana couples of a way to get around the law if they did not want their extra embryos given away to others. The wife could have the embryos implanted, then undergo a traditional abortion, as was her legal right. But being required to take such drastic steps seemed absurd. The woman would have the expense and risk of unnecessary medical interventions (implantation and abortion), all to exercise her right not to procreate.

And now that embryos are being given a right to life, women's abortion rights in general might be in peril. Legislators might say, why is the eight-cell embryo protected and not the eight-week-old fetus? A justice of the Louisiana Supreme Court remarked to me, "The law protecting IVF embryos is a Trojan Horse to get abortion laws changed through the back door."

Artificial wombs would just accelerate the battle. The only barricades against it come from the social discomfort over the "decanting" of children, as it was called in *Brave New World*. In that novel, the government had taken over the control of birthing all children through ectogenesis. All embryos were environmentally engineered throughout development in government-operated artificial wombs. When describing how their prehistory had regarded artificial womb technology, one of Aldous Huxley's government officials remarked: "Take Ectogenesis. Pfitzner and Kawaguchi had got the whole technique worked out. But would the Governments look at it? No. There was something called Christianity."

5

Whose Baby Is It Anyway?

❧

In 1884, Dr. William Pancoast, a medical school professor, was approached by a wealthy Philadelphia couple who had been trying unsuccessfully to have a child. The cause of the problem seemed to be with the husband, so Pancoast looked for someone to donate semen to be injected into the wife's womb. He asked the best-looking member of his class to volunteer, and since the injection was done under anesthesia, Pancoast performed the artificial insemination by donor without the knowledge of either the husband or the wife.

The baby's birth, though, gave the doctor some pause. The infant so resembled the student that Pancoast felt obliged to tell the husband what he had done. The rich Philadelphian, happy to have a child, was delighted by the good doctor's creativity. He asked only that his wife never be told the origin of the child.

A century later, some doctors are still misleading their patients about artificial insemination. A Virginia infertility doctor, Cecil Jacobson, lied to patients about the source of the sperm he was using. Rather than choosing donors to match the husband's characteristics, as he promised couples he would, Jacobson used a source closer to home—himself.

In 1992 federal prosecutors proved through DNA testing that at least fifteen children (and as many as seventy-five) born to Jacobson's patients in the 1970s and 1980s had been created with his own sperm. One couple said they thought something was amiss when they looked at their daughter's first baby picture and saw a striking resemblance to Jacobson.

Dr. Jacobson told another couple that he was using a new technique to impregnate the woman with her husband's sperm. He assured the couple that the husband's sperm was fertile; it just needed a "boost" from a new technique he had perfected. So, at the wife's fertile time, she and the husband went to Jacobson's clinic. The husband produced sperm in the bathroom and handed it to Jacobson. But, later, when the two children were tested, they turned out to be Jacobson's. He had intentionally switched the sperm.

The couple went through months of counseling to determine how to best tell the children about their biological heritage. They did it on a cruise so the children wouldn't associate the news with their home. They revealed that there had been a "mix-up" and the husband was not their biological father, but this in no way changed his love for them. Yet the children were affected; one stopped obeying the dad, saying, "You're not my father."

Other doctors told me—off the record—that in the early days of donor insemination, if a medical student didn't show up on time, sometimes a doctor would use his own sperm to impregnate the patient, excusing himself to the bathroom to masturbate, while the woman lay on the examination table. In Jacobson's case, though, four of his former employees testified at trial that they had never seen any anonymous sperm donors at the clinic.

Jacobson also gave some patients hormone injections that created symptoms mimicking pregnancy, sometimes making them believe for as many as twenty-three weeks that they were with child. He would bring them in for sonograms, claiming to see the baby and check its heartbeat. Later, he would tell the women their "babies" had died and would be reabsorbed into their bodies. He took one woman through three such "pregnancies."

Jacobson was no rogue doctor. A fifty-five-year-old father of eight children with his wife, he was a pioneer in reproductive and genetic technologies, a former professor at George Washington University. He had introduced amniocentesis, the prenatal genetic screening test, into the United States. Couples from around the country flocked to his Reproductive Genetics Center in Vienna, Virginia. A devout Mormon, he would ask his patients to pray with him. During his trial, ninety prominent people wrote letters on his behalf, including U.S. senator Orrin Hatch. Members of the Mormon church fasted on his behalf.

What caused an elite doctor to abuse patients in this way? Prosecutor Randy Bellows speculated that once other doctors learned to do amniocentesis, the prenatal test Jacobson had pioneered, his income dropped and he created false pregnancies to pad his earnings. One patient believed that Jacobson's Mormon heritage allowed him to justify what she called "high-tech polygamy."

But perhaps Jacobson's manipulation of these couples' bodies and lives just came naturally to infertility doctors. Jacobson had said, "God doesn't make babies—I do."

Some experts didn't think his approach was out of line. Dr. Robert Harrison, an Irish IVF doctor who inspected infertility clinics around the world for the World Health Organization, pointed out that Dr. Jacobson's practice was at least as good as the rest—and better than most.

Jacobson's patients felt betrayed, but he felt he had done nothing wrong. Indeed, there were no laws to prohibit doctors from using their own sperm on unwitting patients. Ultimately, the prosecutors charged him with mail fraud and wire fraud for billing patients for sperm he said was from anonymous donors. Which was sort of like getting Al Capone on tax evasion.

"I am in shock, I really am," Dr. Jacobson said when the jury declared him guilty of fraud. "I spent my life trying to help women have children. If I felt I was a criminal or broke the law, I would never have done it." He also said he used his own sperm to protect the patients against getting AIDS.

Judge James Cacheris felt otherwise: "I have not seen a case where there has been this degree of emotional anguish and psychological trauma."

Jacobson protested the guilty verdict, appealing it all the way to the U.S. Supreme Court. He claimed that allowing the couples to testify using pseudonyms, and, in some cases, wearing wigs, violated his constitutional right to confront his accusers. But the appeals courts noted that he and his attorneys knew the identities of the patients; they just weren't allowed to use their names in the open courtroom. The disguises were justified so that the resulting children, who ranged from age four to fourteen, would not learn of their origins through news reports. Jacobson, who faced a maximum 280-year prison sentence, was sentenced to five years in federal prison in Florence, Colorado, beginning in February 1994 and ordered to pay $75,000 in fines and refund $39,205 of the money that patients had paid for his services.

This was not the end of Jacobson's legal troubles. Six couples sued him for child support for the children that he created with his sperm. One couple had been assured that they would get a Jewish donor but had gotten Jacobson instead. Others were promised tall, thin donors; Jacobson is neither.

The doctor's malpractice insurance company, St. Paul Fire and Marine, claimed that it shouldn't have to defend him since it covered only claims arising from his "professional services." The insurer argued that masturbation was not a professional service. That may be, said the federal courts, but insemination was. The insurer settled with the patients for an undisclosed amount.

A prosecutor in the Jacobson case, David Barger, subsequently was assigned to Ken Starr's staff. "I guess I like controversy," he said.

"I guess you are following the sperm trail," I said.

The Jacobson case underscored how lax consumer protections were in the field of artificial insemination. "The history of physician-controlled artificial insemination by donor has been one of shocking neglect of the most rudimentary precautions or record keeping," says Katheryn Katz, professor at Albany Law School.

Most sperm donors are young men, mainly medical students, who do not yet have children themselves. "It's nice pocket change," a medical student told me about the $50 a pop he got for donating. "I can take my wife to a great dinner."

Sometimes, later in life when they become fathers within their marriages, former sperm donors feel remorse about having a child out there in the world whom they will never see. Two psychiatrists I know specialize in treating sperm donors who later experience regret. In Canada, a group of sperm donors banded together to try to find out more about the children they created. Robert Owen, a medical professor at the University of

California, San Francisco, wonders about the children he created in the 1960s as a sperm donor at Harvard. "Were they happy? Were they loved? Were they successful?" he asks himself.

There are no legal limits to the number of children that can be created with donor sperm, nor is there much self-regulation. A survey revealed that 88 percent of practitioners put no limits on how many times sperm donors could be used. In contrast, commercial sperm banks do have limits, but some set those limits quite high, as many as 125 donations per donor.

I interviewed one sperm donor who provided sperm to a bank twice a week, with each donation being split into three or more samples. At the estimated 57 percent success rate for creating children, he could be responsible for 173 children a year.

What happens if two of those children should meet or marry? Already, two doctors have stopped marriages between two individuals they knew to have been fathered by the same donor. But now that the volume of children born through donor insemination has grown dramatically, it is hard to imagine anyone keeping track, particularly as sperm is shipped from city to city. A California doctor I interviewed created thirty-three children while a medical resident at Georgetown. He has provided in his will that if any of the sperm bank babies come calling, they will be allowed to take only $1 from his estate. To prevent intermarriage, he now tells his children by his wife, "Don't marry anyone from D.C."

• • •

The process of artificial insemination is so shrouded in secrecy that few records are kept about donors. When I visited a sperm bank at an Ivy League university, I learned that it never indicated which donor had been used for which woman. If a donor's sperm created a child with a serious disability, there was no way to know which donor it was and stop using him. Since the law of donor insemination was unsettled, the doctors worried that record keeping might later be used to identify a particular donor and seek child support.

In most clinics, doctors exert total control over the process, and the donor choice is entirely within their discretion.

Screening of donors is often lax. The doctors say they rely on medical students' self-reports about family histories of disease or a risky lifestyle. But is it really likely that a medical student, who counts on his professor for a recommendation, will tell the faculty member that he is involved with multiple sexual partners or uses drugs?

A study done at the University of North Carolina found that a majority of sperm donors (including medical students) could not recognize a genetic problem in their family history in order to disclose it. The doctors themselves were not much better at rejecting donors with a questionable history. In a study by the Office of Technology Assessment of the U.S. Congress in 1987, one-third of artificial insemination practitioners were still using fresh rather than frozen sperm, even though fresh inseminations did not provide adequate opportunity for HIV screening. Although approximately half of these doctors said they screened donors for genetic problems, many of them were using inappropriate screening criteria. For example, more than half turned down healthy donors with a family history of hemophilia, even though, hemophilia being an X-linked disorder, such donors couldn't possibly pass on the disease since they did not have it themselves.

In one case, a doctor inseminated a woman with donor sperm

and the resulting child died young of a painful disorder. The couple had a second child, and the doctor used the same donor. The couple had to relive their nightmare when the second child died of the same horrible disease.

Women have contracted gonorrhea, hepatitis B, genital herpes, cytomegalovirus, and other infections from donated sperm. As many as 141 women contracted AIDS, yet the most recent survey, in 1995, found that some clinics *still* did not screen donors for the AIDS virus.

The screening of men donating sperm is more lax than the screening of bulls donating sperm in the cattle industry. Bull sperm is frozen and not used for at least a month after it is collected so that the animal can be monitored to ensure it wasn't harboring an undetected infection at the time the sperm was collected. Records are maintained tracing the health of all offspring of the donor, and the laboratory staff undergo periodic tests of their health to limit the risks of transmitting viruses or bacteria to the samples.

Perhaps its higher quality accounts for it higher price tag. Bull sperm sells for $250 per sample; human sperm goes for an average of $50 per sample.

· · ·

In 1976, a twenty-four-year-old law clerk started giving sperm to the California Cryobank. He filled out a lengthy donor form and disclosed that his aunt had kidney disease. He was not rejected as a donor, though, since, as the clinic doctor examining him later pointed out, 'Nobody's perfect.'

Over the next ten years, donor 276 provided 320 vials of sperm, getting paid $35 per visit. Diane and Ron Johnson chose him because his form said he was athletic with dark hair and eyes, like Ron. They were so pleased with the way their daughter, Brittany, had turned out that they wanted to create another child with the same donor's sperm. But when they called the sperm bank in 1991, they learned that donor 276 had been . . . retired. The clinic had learned that he had the signs of early kidney disease.

When Brittany turned seven, a gym teacher noticed that there was blood in her urine. Subsequent tests revealed cysts on her kidneys and a cyst on her brain. It is likely that the donor had the gene for polycystic kidney disease, which puts his sperm bank children at 50 percent risk of inheriting the illness.

Brittany's parents are now suing the clinic for not protecting them from such a donor. But how realistic are those expectations? There was no genetic test for polycystic kidney disease in 1986 when they used the sperm for Brittany. How much can we expect sperm banks to do to ensure the quality of their donors?

As sperm is treated more and more like a product, people's consumer expectations increase about the resulting children. When David and Stephanie Harnicher sought sperm donation, they wanted to pass the resulting child off as their own. Dr. Ronald Urry suggested using IVF with both sperm from David and sperm from a donor who looked like David. That way, no matter which sperm actually fertilized Stephanie's egg, the child would be viewed by the outside world as a child of the marriage.

Although it used to be common in the early days of artificial insemination by donor, other doctors do not offer this option of sperm mixing. They are concerned that if a man wants his sperm used, too, he may not yet have come to terms with his infertility and be psychologically ready to use a sperm donor.

And they point to evidence that the fertilizing power of the donor sperm is lessened if it is reacting immunologically to the husband's sperm.

The Harnichers settled in on donor 183 who, like David, had dark curly hair and brown eyes. But when the resulting triplets were born, one of them had red hair. DNA tests revealed that donor 83, rather than 183, was the father. Donor 83, who had actually been on the Harnichers' short list of four donor finalists, had straight auburn hair and green eyes.

Other couples might have been happy to have three healthy kids. Instead, the Harnichers sued the clinic for $500,000, saying they had "suffered severe anxiety, depression, grief and other mental and emotional suffering and distress which has adversely affected their relationship with the children and with each other." In fact, the couple divorced—they said due to the stress of the sperm mix-up—six months after the children were born.

Stephanie testified at trial that she could say "with probability" that children of donor 183 would have been more attractive than her children, even though she had never seen either donor 83 or donor 183.

Since there was no way to prove that the Harnichers would have had any children, let alone more attractive children, if they had gotten the donor of their dreams, the Supreme Court of Utah refused to speculate on the road not taken. Their 1998 decision was a close one, though; three justices ruled against the couple, but two would have ruled in their favor. The majority said that "destruction of a fiction" is not grounds for a lawsuit. Dissenting Justice Christine Durham held that a woman's right to choose includes the right to choose to have children that she can believe are her husband's. If the clinic had done its job, she said, "the 'fiction' would never have been labeled a fiction; it would simply have been an 'alternative reality' for the family."

. . .

Although artificial insemination has been used in the United States for the past century, no one yet has studied its impact. There is no social term for the relationship between the donor and the recipient, even though they are collaborating in the intimate act of creating a child.

The courts, too, have been befuddled by these arrangements. One of the first artificial insemination cases in the country, in Illinois in 1954, ruled that artificial insemination by donor was adultery, even if the husband consented. A later California case disagreed with that decision: "Since the doctor may be a woman, or the husband himself may administer the insemination by a syringe, this is patently absurd; to consider it an act of adultery with the donor, who at the time of insemination may be a thousand miles away or may even be dead, is equally absurd."

While such cases settled the question of whether the anonymous sperm donor should be considered the father, a new set of legal questions arose when unmarried women began to seek access to donor sperm. A national survey of infertility doctors in 1979 found that fewer than 10 percent would provide sperm to single women.

In 1980, a thirty-six-year-old woman, Mary Ann Smedes, sued Wayne State University's infertility clinic because it restricted artificial insemination to married couples. With the American Civil Liberties Union representing her, she challenged the clinic's policy, saying it violated her constitutional reproductive freedom and her right to equal treatment.

She said in her affidavit that although she was interested in remarrying, she felt she must bear a child in the next year or so because of her age. She pointed out that whether she remarried or not, she felt competent to be a parent and provide for all of her child's emotional, financial, and social needs. "Any woman could get pregnant by a man if she wanted to," she pointed out. "With donor insemination, there is less stigma for the child."

The case never got to court. It was settled with the university's agreement to drop its marriage requirement and to consider the woman for its artificial insemination program.

When I called to congratulate Mary Ann Smedes on her legal victory, I learned that she had not gone ahead with the insemination. Over the course of the lawsuit, she had received more than two hundred threatening letters from people opposed to her single motherhood. And while the doctors had agreed they wouldn't reject her because of her marital status, they discouraged her on other grounds, mainly her age.

The search by single women for a sperm bank that would serve them led in 1982 to the creation of the Sperm Bank of California, which caters primarily to unmarried heterosexual women and lesbian clients. Women all over the country can order an "overnight male"—sperm delivered to their door via FedEx. When the liquid nitrogen tank arrives, they can take it to their gynecologist for insemination or do it themselves at home with a turkey baster.

Obtaining sperm is about to get even easier. Xytex Corporation, a Georgia sperm bank, is selling franchises to drugstores. Xytex will put a liquid nitrogen tank in pharmacies and pay the store $50 for each $175 sperm sample sold. "Anywhere in a large metropolitan area, by nature 20 percent of the population trying to have a child is infertile," says

David Towles, director of public relations for Xytex. He tells pharmacies they can break even as long as they can dispense about eight samples a month on doctors' orders.

I try to imagine a future where women run into a drugstore to pick up some sperm and a package of Marlboros. "No," she will say. "This one's special. Make it Dunhills."

*   *   *

Elise Green had a hysterectomy in her twenties, so her significant other, Shari Wilson, was artificially inseminated. But when they broke up before the child was two, Wilson tried to keep Green from visiting the child. Their pending case is one of more than two dozen lesbian-parent visitation suits. Half the time, courts refuse to recognize the non–birth parent's link to the child and deny her visitation rights.

"It's a tragedy not just for the toddler who lost a parent and doesn't know why," says Kate Kendell, executive director of the National Center for Lesbian Rights. "It's also a tragedy for the larger lesbian community, having a lesbian go to court and insist the law not respect her family."

Kendell chastises women for blocking their former partners' access to the children. "Heterosexuals don't get it both ways. Neither should we. Even if they are angry or feel betrayed, both members of the couple should be allowed access to the child."

Lesbian parenthood through artificial insemination has some unusual twists to it. On rare occasions, both members of the couple get pregnant simultaneously with donor sperm to further share the bond of parenthood. More commonly, one has a greater interest in bearing children; the other may even be infertile.

Some lesbian couples seek donors from among their gay

friends. A lesbian couple in Chicago, Lynn Alleruzzo and Charlene Crotty, used Kevin Green's sperm to create their child. Green (no relation to Elise) helped in the pregnancy and delivery, but when the women refused to let him continue seeing the child, he sued for joint custody. In 1998 they reached a settlement, giving Green and his male partner the right to have contact with the child, leading to a situation where the child will have two moms and two dads.

Lesbian women who wish to avoid later custody disputes are turning to even more complicated technologies. Some use donor sperm to fertilize one of the women's eggs and then implant the resulting embryo in the other woman. In San Francisco, a judge scrutinizing such a relationship allowed both women to be listed on the birth certificate as parents.

"That judge had presided over hundreds of lesbian stepparent adoptions," says Kendell, who points out that other judges might not be so open-minded. "If you tried such an arrangement in Des Moines, you'd be picking the judge up from the floor. Then some legislator would be introducing a bill to ban it."

Another set of individuals is trying to gain rights to participate in artificial insemination. Fourteen men on California's death row have made a claim—so far rejected by the courts—that their reproductive rights include the right to store sperm so that they can later (maybe even after their deaths) become fathers. Sadly, five of the fourteen are on death row for killing kids.

• • •

The Sperm Bank of California has redesigned what a sperm bank can be. The women who came to it were interested in a wider range of donors. One said to me, "I hate doctors, so why would I want a medical student to father my child?"

The clinic is unique in asking each donor whether he is willing to be identified to the resulting child when he or she reaches age eighteen. More than forty percent say they would. Some prepare scrapbooks or videotapes that can be given to the child to fill in information about the donor's side of the family.

Some countries go even farther. In Sweden, Germany, and Austria, the children created through donor insemination have the right to learn the donor's name.

Children of donor insemination in the United States do not have such a legal right, though. Some long for information about their biological fathers. Twelve-year-old twins born with anonymous donor sperm made a plea on national television to meet their "dad." The sperm bank pressured him to talk to the children, which he did, even though he felt angry and betrayed since this hadn't been part of the original deal.

At age thirty-one, Suzanne Rubin learned that the man she had called Daddy all her life was not her biological father. Her mother had undergone artificial insemination with sperm that she had been told was donated by a Jewish medical student from the University of Southern California.

"It took away half of what I was and replaced it with a blank space," Rubin told me. She was determined to fill in the space.

Posing as a student writing a term paper, Rubin gained access to the University of Southern California archives and pulled the records of four hundred men who were medical students, residents, or interns in the year her mom was inseminated. She made a list of those whose names sounded Jewish. "The German names were the toughest," she says. "If I couldn't tell whether a name was Jewish or not, I went to the marriage records for the Los Angeles area." If the man had been married in a synagogue, he stayed in Rubin's active file. Parenting possibilities: fifty-five.

Rubin next consulted a geneticist, who told her that since she had red hair and blue eyes and her mother had black hair and dark brown eyes, there was a 99 percent probability that the donor would not have dark hair and dark eyes. She and three friends separately went through photos of the remaining men, either in student files or in medical directories. If eye color was too difficult to determine, Rubin used driver's license information. She and her friends were in agreement. They each identified the same ten men as potential fathers.

Then Rubin ran across a photo of her mother's doctor. "He looks exactly like me," she asserts. An intermediary sent that photo and Rubin's photo to the doctor with a note, "We've noticed the resemblance."

He didn't write back.

6

Cybermate

~

I am letting my fingers do the walking, searching for a sperm donor on the Internet. My first stop is the Web site for OPTIONS, a baby brokerage firm that puts infertile couples in touch with sperm and egg donors. When you enter its Web site, you learn that it offers "personal support." The introductory page says, "We know that sometimes a few encouraging words, or just a hand to hold, can make all the difference in the world."

I wonder how anyone will hold my hand in virtual reality. There is no hand-holding icon or I-feel-your-pain icon. All I see is storks.

I access the sample sperm donor profile, number 1049. The number is supposed to make you feel comfortable, as if it guarantees over 1,000 donors in stock. But there is nothing to prevent a donor service from beginning its numbering at 1045. One sperm bank I visited has infertile couples agonizing over choices in a ten-page questionnaire about what they want in a donor. Dimples or not? Jewish or Catholic? Musical or math ability? The endless array of questions made it seem as if, behind the Wizard of Oz door, there would be hundreds, if not thousands, of donors to be able to fulfill every nuance of the couples' desires. In truth, there were just twenty.

Number 1049's photo shows a clean-cut, cute Californian. I continue browsing his five-page donor profile. He is a member of the Clean Oceans Campaign and the Surfrider Foundation. He describes himself as "secure, sensitive, innovative, intelligent, creative, thoughtful, ambitious, competitive, respectful, comedic, and optimistic." SAT score 1355; not bad. His fifty-four-year-old mother is an intelligent and adventurous painter, healthy, except that she wears reading glasses. His brother is a "developer."

While I could buy sperm for a couple hundred dollars from a local sperm bank, it will cost me $2,370 (plus medical fees and screening fees, plus a surcharge if I live outside California) for one of the OPTIONS donors.

Pacific Reproductive Services, another Internet sperm provider, has seen its business increase 25 percent since it went on-line. It now ships sperm around the globe, with orders from Japan, Europe, and Israel. But the dozens of sperm banks offering this new export are raising concerns. Is this a new form of imperialism, with American gametes seeding the globe? And who is to blame if something goes wrong? In England, the Human Fertilisation and Embryology Authority has issued a warning about buying sperm over the Internet, since there is no certainty that donors have been properly screened.

• • •

When doctors chose sperm donors, couples were only given very general information about what the donor looked like. In a study by Betty Orlandino, sperm recipients differed markedly from their spouses in how they pictured the donor. "The wives thought the donor probably looked like Robert Redford," says Orlandino, "while the husbands pictured him looking like a Skid Row bum."

The Internet connection adds a new dimension to artificial insemination: the possibility of viewing a photo of the donor and, in some instances, listening to a tape of his voice. At Southern California Cryobank, many donors are aspiring L.A. actors who make ends meet by selling sperm. How will an infertile husband feel when his wife decides to use the young, handsome donor 1049? Will the wife fantasize about the donor in his wetsuit? What happens if she runs into him in the grocery store?

These digital dads seem more like part of a dating service than a medical clinic. In fact, one Israeli company matching sperm donors and recipients entered into the field with its New Life Web site after its success with Net Match, a service for connecting singles.

Throughout the on-line procreative personals, appearance is stressed. A typical donor profile at the Pacific Reproductive Services Web site says, "angular Nordic features, handsome eyes." Another says, "cute rounded face, nice eyes, well balanced features." That latter phrase makes me wonder about the other donors. If "balanced" isn't specifically listed, will I end up with a child who looks like a Cubist painting?

The on-line donor sperm catalog for the University of Illinois at Chicago lists the donors' heights, which most clinics no longer do. A New York City sperm bank director told me that no one requested shorter donors. I notice that more than 90 percent of the donors in the Illinois catalog are over five-foot-ten, including some that top the charts at six-foot-six and six-foot-seven.

As I surf the Web, I come upon a service, www.egg
donation.com, that is recruiting women to share eggs with
infertile couples. There are more than seventy-five such egg
donation programs in the United States.

The prices are enticing. When egg donation began in 1984,
Richard Seed paid donors just $250. By 1994 the going rate
was $1,500. In 1998 St. Barnabas Hospital in Livingston, New
Jersey, boosted its rate from $2,500 to $5,000 during an egg
donation bidding war. But the largest fee—$35,000—is being
offered by an anonymous couple who specifically want an attractive, intelligent, *Princeton* woman's egg.

The soothing digital directive makes potential egg donors
feel that they will be safe and secure as if they were creating
a virtual pregnancy. When one potential applicant called a
Pennsylvania clinic and asked about the risks of donation, she
was told there were none. In truth, the risks are plentiful, and
their magnitude is unknown.

The application process itself can be unsettling as four out of
five women are rejected. Most forms ask a woman to disclose her
weight, which, apparently, is as key to potential recipients as is
height for male sperm donors. Maureen Mendick of Raleigh,
North Carolina, chose to be an egg donor when she learned that
a number of friends were going through infertility. She didn't
need the $1,500; she and her husband are both well-paid engi-
neers. But as the mother of a one-year-old, she wanted to help
another woman become a mother as well. After undergoing a
medical exam and psychological testing, she was ushered into
the doctor's office, where he looked her over. "When I look at a
donor, I look to see if I would want to raise that person's genetic
material," he told her. "And when I look at you, I can honestly
say yes."

My friend, law professor R. Alta Charo, did not fare as well when she applied to a surrogate mother program to see what the process was like. She has crossed eyes and is not model thin. The clinic rejected her, implying, who would want a child like you? "I was not commercially viable," says Alta.

Once a woman is chosen, she faces a grueling medical regime. She injects fertility drugs into herself daily for two weeks. The painful shots cause the woman's moods to swing widely. (One infertility doctor calls it "the Hitler-Bambi syndrome.") She may experience headaches, hair loss, cramps, and nausea. She bloats up, and risks kidney or heart failure if she is not properly monitored. (By contrast, a sperm donor spends a few minutes in a room that looks like a high school boy's fantasy. It has soft- and hard-core magazines, a video player with X-rated tapes, and—essential in this business—a vinyl couch.)

The monitoring of egg donors involves daily visits to the clinic for blood tests and ultrasound screening. When the ovaries are ripe, as many as several dozen eggs are removed with a fifteen-inch-long needle inserted through the vagina. Either a local or general anesthesia is used, subjecting the donor to those risks as well. According to some studies, she also has a greater chance of getting ovarian cancer later in life.

For an increasing number of young women, particularly those who aren't informed of the risks, the money can be enticing. Russian and eastern European women—a prime source for nannies a few years ago—are now providing prenatal child services by donating eggs. Job Track, a service at the University of California at San Diego, puts students in touch with egg brokers; some women use the money to pay their tuition. Carrie Specht, while a graduate student at New York University's film school, used the $9,000 she made from four egg donations to make two short films she showed at the Sundance Film Festival. She named her film company Zygot Productions (after the word *zygote*, the scientific term for a fertilized egg).

For some egg donors participating in PC parenthood, the process is not as fulfilling as was expected. Half of anonymous donors have second thoughts about their participation—including concerns about what life would be like for the resulting child. The whole experience, says Maureen Mendick, made her feel as if she were a factory.

Six months after her couple gave birth to a child, she learned they wanted her to donate eggs again to create a sibling. Maureen wondered why they were requesting more eggs, since the doctors had harvested nineteen.

She had assumed she was helping out one couple, but now she wonders if her nineteen eggs were split among three or four couples, each paying the donor fee. She was troubled about having her biological children split among several families, without either her or the families knowing about it. "That's the problem with secrecy," she says. "You can't really check on what the clinic is doing."

Part of the problem, observes University of Pennsylvania bioethicist Arthur Caplan, is calling the procedure a donation in the first place. These are vendors, not donors. "You might get further in protecting the egg donors if you said this is a market and you need legal representation," he says.

Because of the risks presented to women who are recruited to be egg donors, some clinics prefer to use their own IVF patients. According to a 1992 survey, almost half of IVF clinics use their patients as egg donors. Since these women are already taking fertility drugs, they seem to face few additional physical risks by donating. Some clinics even offer IVF to women at a reduced cost if they are willing to give some of their eggs to other women. Yet that seems to be putting the women in the position of a tragic Sophie's Choice. In order to be able to create a child for herself, she has to give up a potential child to someone else.

Even when there is no systematic policy for egg donation, some clinics do nothing to prevent patients from being solicited by other patients. Sometimes, an infertile woman who cannot produce eggs will show up on an egg retrieval day and ask an IVF patient to share some of hers. Such personal pressure may be hard to resist.

The IVF patient's own needs may not be well served when she acts as a donor as well. The temptation for the doctor will be to pump her up even further with hormones to ensure that she produces dozens of eggs for donation. The IVF patient might not get pregnant, while the recipient does. In fact, that is exactly what happened in the first case of egg donation, in Australia. One of these days, a frustrated, childless IVF patient is going to sue a recipient of her eggs for visitation rights to the children that have been created.

Egg donors have started advertising directly. One such ad reads, "JEWISH EGG DONOR. . . . Very attractive, in perfect health. Willing donor for couple seeking great genes." I read the ad several times. After all the tragedy that had come to Jews in the name of eugenics and racial purity, it disturbed me that a Jewish woman would suggest that childbearing choices should gravitate toward great genes.

At the Web site of the Center for Surrogate Parenting and Egg Donation in Beverly Hills, there are three hundred egg donors available. Each fills out forms listing her favorite books and movies, her weight, her own and her parents' health history, and her "philosophy of life." One donor attracts couples who share her philosophy that "what goes around comes around."

These data, along with photos, are how couples choose a potential donor. The cost of a match: $6,000 up front, and all the medical expense of the transfer on top of that.

The shortage of egg donors in countries where payment to them is banned leads many Europeans to order eggs from the Beverly Hills center. An English couple, for example, sent the husband's sperm to California, where it inseminated a donated egg, resulting in an embryo that was express-mailed back to the couple.

A British talk show host asked twenty-year-old Louise Brown, the world's first in vitro fertilization child, "How would you feel if you learned you had been sent to your parents in a jiffy pack?"

"I would be disgusted," she replied.

Louise is proud of her own beginnings, yet someone might just as easily have asked her, "How would you feel if you learned that you'd started your life in a small plastic dish after your father masturbated in the next room?"

That's the problem with the public level of discussion of reproductive technologies. No one is giving sufficient thought to what criteria should be used to compare technologies and make decisions about what should be allowed in the first place.

• • •

The scarcity of eggs—as compared to sperm—has led to heated debates over who should be chosen as a recipient. Egg donation could revolutionize parenthood by letting postmenopausal women achieve a pregnancy with hormones and a younger woman's eggs.

Yet when I appeared on the *Newshour with Jim Lehrer* with an infertile woman who had two sons through IVF, she told me that donor eggs should not be used for postmenopausal women because that was not "natural." Just a few years earlier, the same argument had been made against IVF itself.

To women who are past menopause, donated eggs can be a fountain of youth. In November 1996, a sixty-three-year-old woman named Arceli Keh gave birth, after lying about her age to a California infertility specialist and obtaining eggs from a younger donor.

Another woman, sixty-two, chose IVF with donor eggs after her nineteen-year-old was killed in a car accident. An Italian pioneer in infertility, Dr. Severino Antinori, has helped seventy postmenopausal women become pregnant. The pregnancy success rate for older women using younger women's eggs is 25 to 35 percent, which is even higher than the success rate for younger women whose problem is infertility, rather than menopause.

Yet female television producers who asked me on their shows to talk about postmenopausal pregnancy were vehemently opposed to the process. I tried to figure out why and can only surmise that it was because they had finally made their peace with their biological clocks, deciding to have or not have children by the time they reached their late forties. Egg donation to women past menopause, like the endless streams of new treatments offered to young infertile women, meant that women can never come to closure on the question of the relative merits of motherhood versus career.

Dr. Zev Rosenwaks, director of the Center for Reproductive Medicine at New York Hospital–Cornell Medical Center, is opposed to providing eggs for postmenopausal women. He argues that the situation is not comparable to Senator Strom Thurmond's fathering a child at seventy or actor Anthony Quinn creating his eleventh child at age seventy-eight.

"Since men do not undergo pregnancy, there is no chance their baby's health will be compromised," says Rosenwaks. But since few people have tried postmenopausal pregnancy, no one really knows what the risks will be. And it is ironic that Rosenwaks's clinic is willing to try a score of other types of interventions: in vitro fertilization, embryo freezing, preimplantation screening of embryos, injection of sperm into eggs, and virtually every other trendy new technique, without an equal concern for the unknown risks to the child.

This is rationing medical services by age, says University of Minnesota law professor Susan Wolf, who compares it to setting an age maximum on who can receive a kidney. "Rationing should not be done by individual physicians in an ad hoc fashion: the social implications are too great."

In its guidelines discouraging egg donation to postmenopausal women, the American Society of Reproductive Medicine says, "Just as oocyte donation to prepubertal girls is unacceptable, so should it be unacceptable for postmenopausal women to bear children." It seems insulting to older women to say that putting an egg into a ten-year-old girl, before she is competent to consent, is the moral equivalent to getting a fifty-year-old pregnant at her request.

ASRM also says it wants to protect older women against the risks to their health by preventing them from getting pregnant. But even for younger women, being pregnant is riskier than not getting pregnant. Maybe we should just view egg donation as the reproductive Viagra for older women—risky but enticing.

"The limit is not age; the limit is physical," says Dr.

Antinori. "What I want to see is a law that controls the centers, and not the wishes of the people."

Jonie Mosby Mitchell, the mother of four children in her first marriage, used reproductive technology to have a child at age fifty-two. One of her daughters was pregnant at the same time she was. Her oldest daughter was annoyed. "Mom, if you've got extra time," she said, "spend it with *my* kids."

Egg donation has also led to some interesting linguistic problems. At one academic meeting, a questioner asked, "What do you call a woman who provides the genetic, but not the gestational, component for reproduction?"

Sociologist Barbara Katz Rothman was swift with her reply: "A father."

7

Wombs for Rent

❧

The practice of surrogate motherhood is as old as the Bible, in which the sterile Sarah instructed her husband Abraham to impregnate her handmaid, Hagar. In the 1980s, though, it was rediscovered with a vengeance by talk show hosts. At first, surrogacy was a blue-collar affair, with women of modest means altruistically having babies for other such women. But once it was clear that it worked, that women *could* give up a child to an infertile couple, the process became gentrified. Lawyers began charging $10,000 to $30,000 to act as matchmakers between surrogates and well-off couples. They began to offer a fee to the surrogates as well.

In 1980 Carol Pavek, a midwife in Amarillo, Texas, decided to have a baby for a California couple, Andy and Nancy. The Californians took the tiresome three-and-a-half-day bus ride to the Texas panhandle to meet Carol and her husband, Rick, face-to-face. The trip left them totally exhausted, and the excitement and novelty of meeting the woman who might bear their child put them emotionally on edge. When they walked into the Pavek home, Nancy burst into tears.

Carol and Rick immediately took to the couple. "They were blue-jean/T-shirt/fast-food-type people," says Carol. The two couples were similar in personalities, financial worries, and attitudes toward children. Andy had red hair and nearsightedness, like Carol, and admitted that under his beard "there is a terrible receding chin."

"Great," said Carol, "if the baby gets a receding chin, we can blame it on you instead of me."

Early in their visit, Carol took Rick out of earshot. "What do you think?" she asked.

"I think they're wonderful," he replied.

So they went back into the living room and made this offer: "Why don't you stay in our home with us? If you save the money you would have spent on a hotel room, you could afford to go back by plane rather than by bus."

During the pregnancy, the couples had to limit their phone calls back and forth, because of the cost. But there was still a lot of contact—at least a letter a week. As the pregnancy progressed, the two families began to ponder what would happen if there was a problem in the complex legal proceedings that were to take place in Texas and California. "Everyone agreed that we would never put the baby in foster care," says Carol. "Either we would keep the baby, or they would keep the baby."

Some people were critical of Carol's decision to be a surrogate, likening it to adultery. "You cannot commit adultery with a syringe," replied Carol. "Yes, my body is being used. When I die, my eyes are going to be used. My kidneys will be used. It is not my body in the first place. I come from a Christian philosophy which says we are our souls, the body is merely a vehicle we use while we are here on this planet. We shouldn't be so possessive about our bodies."

Carol also received letters, some supportive, some negative, some just plain weird. A nurse was irate that Carol had inseminated herself under unsterile circumstances; the nurse pointed out that even cattle breeders use sterile circumstances to inseminate their cattle. "I was tempted to write her back to ask how long her husband boiled his penis before he inserted it," Carol said.

When the baby was born, the doctor said that he wanted to lay the baby on Carol's stomach. "Have you forgotten, this isn't my baby?" she asked. She was worried that she might bond with the baby—which, after coming so far to have a baby for someone else, she didn't want to do.

"But, Carol," he continued, "it's the only warm place in the room."

The baby was gently lowered onto Carol's abdomen. She slowly opened her eyes and was relieved to find that she didn't have any feeling of possession. Her only thought was *what a gorgeous baby*.

When Carol returned home from the hospital, bags of letters from other couples seeking her services had arrived. She was overwhelmed by the sadness they represented. She began to look through them to see if there were any couples that she might feel as close to as she had to Andy and Nancy. She was able to rule out about two hundred couples fairly easily. "I want to feel that I've personally selected a good home for my baby," she thought at the time. "I don't want to worry for the next eighteen years whether the couple is financially and emotionally stable."

Carol refused to serve as a surrogate for a couple who wanted a child just because they needed an heir, and for a couple who she thought would not spend enough recreational time with the child. In making her decision, Carol tried to picture the type of life the child would have. One couple, she believed, did not so much want a child as someone to pass their corporation on to. "I could see him in the Ivy League, maybe even in military school from the time he was six years old," envisioned Carol. "The couple didn't seem to have any warm feelings of 'let's go make mud pies in the rain.'"

By the time she was considering having her second surrogacy child, the process had changed and surrogates were being paid $10,000, with another $10,000 going to the attorney.

Carol had reservations about the men who were now seeking surrogates. "The men who would be the best fathers," she said at the time, "are men who do not mind being spit up on, who will sit on the floor with the baby. They are not the aggressive, assertive attorneys or doctors who are the ones who were most likely to be able to track me down or who are the ones who could afford to pay a surrogate. The men who would be the best fathers might not be able to pay."

The mental impasse that daunted Carol about whether or not to accept payment was overcome by a physical accident. One day, Carol was backing her old green Opal out of the drive at the precise instant that the neighbor across the street was backing out of her own driveway. Both cars were wrecked. *Well, I guess, yes, we will accept money,* Carol thought.

There was a complex interweaving of motives in Carol's decision to go on to her second surrogate pregnancy. The articulate, feminist Carol talked persuasively about women's reproductive freedom and how surrogate motherhood was a pillar of that freedom. But a quiet, much younger-sounding voice from within her spoke almost wistfully about what the money would mean. "Just once, I would like to be able to go into a car dealership, be well treated, and pick out a car without anyone quizzing me about how I would ever be able to afford it," she said. These sides came together as she telephoned Michigan attorney Noel Keane, the man who first introduced the idea of wombs for rent.

Once surrogacy moved into the paid realm, it attracted the attention of feminists concerned about the exploitation of women. By the late 1980s, many feminists were frustrated that women had hit a glass ceiling in professional jobs and been virtually excluded from political office. They shuddered at the idea that just because a man had mobile sperm and a checkbook, he could buy a woman's reproductive power, without having to enter into any intimate relationship with her. On the other hand, feminists had always argued that a woman should be able to make her own decisions about her body. Some would choose to mother, others to abort, and perhaps others to have a child for someone else.

At a conference of law professors I attended, it was suggested that surrogacy was wrong because women's boyfriends might talk them into being surrogates and because women might be surrogates for financial reasons. But women's boyfriends also might talk them into having abortions, and women also might have abortions for financial reasons. The fact that a woman's decision could be influenced by the individual men in her life, by male-dominated society, or by economic circumstances does not by itself provide an adequate reason to ban surrogacy any more than it does to ban abortion.

But the surrogates I met did not seem to have been coerced. Surrogate mother Donna Regan testified in New York that her will was not overborne in the surrogacy context: "No one came to ask me to be a surrogate mother. I went to them and asked them to allow me to be a surrogate mother. I find it extremely insulting that there are people saying that, as a woman, I cannot make an informed choice about a pregnancy that I carry." Regan pointed out that she, like everyone, "makes other difficult choices in her life."

"Surrogacy does present potential psychological and physical risks to the women involved," says surrogate mother Jan Sutton. "But generally our society has allowed people to undertake potentially risky activities so long as they have given voluntary informed consent. Men are allowed to be coal miners, firefighters, race car drivers, and U.S. presidents for compensation even though these occupations involve risk. It would be inconsistent and demeaning to women to deny them the chance to be paid surrogate mothers."

The *Baby M* legal proceedings, in which Mary Beth Whitehead changed her mind and decided to keep the child she had contracted to bear, clouded, rather than illuminated, the issues surrounding surrogacy. Both sides hired experts to prove the other would be unfit parents. There was no hard evidence that Mary Beth or Rick Whitehead or William or Elizabeth Stern would be unfit parents; none was thought likely to abuse or neglect Baby M. So the experts undertook the psychological equivalent of an electrocardiogram to determine if there were any blips in their personalities that would make them less worthy to parent Baby M.

The results were terrifying to any parent or potential parent. A psychiatrist, psychologist, and social worker visited the immaculate Whitehead home, where various of Baby M's possessions—panda bears, a crib—waited forlornly for her return. Under the watchful gazes of the mental health professionals, Mary Beth bathed her daughter, fed her a bottle, played patty-cake with her, let her wander the living room in a walker, and rocked her to sleep. It might have seemed like the picture of maternal-child bonding and bliss. But when the experts rated her mothering based on these events, they managed to find fault in virtually everything she did.

One of the expert witnesses, Dr. Marshall Schechter, criticized Mary Beth for letting the baby play with stuffed animals instead of pots, pans, and spoons. He also faulted her for saying "hooray" instead of "patty-cake" to reinforce the baby when playing patty-cake. Mary Beth was also criticized, by other experts, for her lack of openness to mental health counseling—and for dyeing her hair.

Feminists naturally flocked to Mary Beth's aid. But then they went farther. Rather than just helping surrogates who had changed their minds, they started pressuring other surrogates to keep the children they had contracted to bear.

Cynthia Custer was a surrogate in a Maryland program in which the couple and the surrogate remained unknown to each other. Near the end of her pregnancy, Custer decided she wanted information about the couple who would be rearing the child. When the director of the program refused to tell her, Custer took her story to the *Washington Post*.

When the article appeared, says Custer, Sharon Huddle, a feminist lawyer who was a member of a group called the National Coalition Against Surrogacy, told her, "This baby wouldn't want to be given up. He doesn't have a choice." Then she pressured, in a higher voice, "Please, Mommy, don't give me up."

The *Baby M* case was generally described as a battle between the classes. Much was made of the fact that the Sterns earned more money than the Whiteheads. But, then again, the Sterns were older and farther along in their careers. Moreover, William

Stern was not from a privileged family, nor was Mary Beth Whitehead from a lower-class one. Mary Beth's father had a master's degree and taught school. Bill's father worked as a short-order cook. Rather than acknowledge that a woman might become a surrogate willingly and enthusiastically, certain observers constructed a fable about a desperate woman needing money.

And focusing on a salary discrepancy ignored the fact that Mary Beth Whitehead was not acting with the baby's best interest in mind. "I gave her life and I can take it away," she said, threatening to kill the infant rather than give her back to the Sterns.

Payment for parenting strikes some people as unseemly. Yet most people agree that child rearing is a more important component of parenting than childbearing, and society allows all sorts of payment to surrogate child rearers, such as baby-sitters, nannies, and day care centers.

One surrogate mother pointed out to me that she received $10,000 (of which she paid $3,000 in taxes) for being a surrogate, turning the child over to the biological father. Yet she received much more criticism than if she had received a tax-free $10,000 for "reasonable expenses" and given the child up through a private adoption, turning the child over to a stranger.

The biological mother in the surrogacy situation seeks out the opportunity to carry a child that would not exist were it not for the couple's desire to create a child as a part of their relationship. She makes her decision in advance of pregnancy for internal reasons, not external forces. While 75 percent of the biological mothers who give up a child for adoption later change their minds, only around 1 percent of the surrogates have similar changes of heart.

In adoption, if the biological mother changes her mind, the child goes home with her because the adopting parents have no legal claim. With surrogacy, though, the father who provided the sperm (and in cases of IVF surrogacy, his wife, who provided the egg) does have a legal claim to the child. Giving the surrogate a right to change her mind and assert her parental rights thrusts the baby into legal limbo, with years of litigation to determine who will raise the child.

Nadine Taub, a professor at Rutgers University, tried to mediate between the various feminist positions by asking activists from around the country, including me, to join a committee on "Reproductive Laws for the 1990s." I, in turn, began visiting surrogate mothers around the country to find out how these women were faring. Over the next few years, I interviewed more than eighty surrogates and couples, and visited surrogacy centers in five states.

I admit that I went into the process looking for evidence that the women had been exploited. I recall driving up to the home of a Latina surrogate in Southern California, a tiny speck of a house in a rundown, gang-ridden area, and thinking, "Finally, I've found an example of a woman being taken advantage of." In our interview, though, she surprised me with her decisiveness and commitment. She told me, "I can't feed the hungry, I can't stop war, but I can change the world in a small way by helping a couple become a family."

She had two young children of her own and wanted to assure me that they did not fear they were in danger of being given up. So she put a photo of the couple on the refrigerator door and said, "This is Mary and Tom and Mary's tummy is broken, so I'm having this baby for them."

Among surrogates, I had expected to find women who put

little value on motherhood; what else could explain their willingness to get pregnant and then turn the child over to someone else? Instead, I found women for whom motherhood was a central focus of their lives. They couldn't imagine life without children and thus felt empathy for the infertile couple. Several other studies of surrogates found their psychological profiles were not that different from other women their age, except that they showed a slightly higher degree of altruism and risk taking and an easier time with pregnancies. And the surrogates' own children I spoke with did not have fears they would be given up because they felt—as most of the surrogates felt—that the baby belonged to the couple from the moment of conception.

The infertile couples I met were baby boomers who were not willing to take no for an answer when it came to having a child. "We're the generation that has discovered technology," John (not his real name) told me. "We're the generation that has the biological knowledge to apply the scientific method in dealing with these issues. We're not willing to accept infertility without looking for a solution."

The solution John and his wife, Elizabeth, found was surrogate mother Jan Sutton. Jan wasn't afraid of new experiences; she had always volunteered for experiments when she was getting her degree at the University of Michigan. In pursuing surrogacy, she did not need the money. She was a well-paid pediatric intensive-care nurse in San Diego, and her husband also had a good income.

At Jan Sutton's fertile time, Elizabeth flew down from northern California with her husband's sperm sample in a lunch sack. Before her departure, she refused to put the bag through the airport X-ray machine.

"What's in it?" asked the guard, unfolding the edges of the bag.

"It's my future child," she said.

He refolded the edges and handed it back to her without a word.

In San Diego, Jan met her at the airport. "This is a little odd," Elizabeth said. "This is not how I pictured the creation of my child."

It was odd for John as well. "Here was this very attractive, very intelligent woman that under vastly different circumstances would have been somebody I would have dated—and she was bearing my child," says John. "I had to detach from that. It's not that I wanted to have an affair or anything. I have a wonderful wife. I am happily married. I don't want anything mucking that up. It's just that way down in the older brain, where sex and reproduction are connected, we get into some pretty delicate territory."

During the pregnancy, Jan and her children grew close to Elizabeth and John. Because they lived in different parts of the state, they joked about taking a joint vacation, along with the baby, in order to spend some uninterrupted time together.

"Why not a camping trip?" suggested John. So, Mother's Day 1986 found a most unusual family together in Yosemite. The fifteen-month-old baby was the focus of attention of her father, her half-siblings, and, of course, her two mothers.

This was only the second time that Jan had seen the infant since giving birth at the hospital. Jan took a deep breath and moved toward the child. She was a round-faced, honey-colored blonde, closely resembling Jan's own daughter, Kris. Jan smiled—and the little girl ignored her. Jan realized that she had no desire to take the baby home, nor did the baby show any desire to be with her.

Not all surrogacy arrangements worked out that well, though. Surrogates were such a gold mine for attorneys that they were reluctant to turn away women even if they, or the couple, might be harmed by the process. Noel Keane used as a surrogate a woman who extorted money from the couple and a woman who drank heavily, producing a child with fetal alcohol syndrome.

Part of the reason that Noel Keane's arrangements went awry was that he did not take seriously the idea of screening. Philip Parker, the psychiatrist working with Neal Keane's Michigan program, told me that "I don't screen any women out. I wouldn't know what a good surrogate is. For all we know, the more disturbed she is, the better."

In the infamous *Baby M* case, psychologist Joan Einwohner in Noel Keane's New York office had interviewed Mary Beth Whitehead when she applied to be a surrogate. Einwohner recommended that, before Mary Beth was allowed into the program, more in-depth screening and counseling be done to determine whether she would be able to relinquish the child in the final analysis. "She may have more needs to have another child than she is admitting," the psychologist wrote in her report. Yet Keane still hired Whitehead, without telling the Sterns or her about the troubling report.

In another of Keane's cases, the problem was not that too many parents ultimately wanted the baby, but too few. Judy Stiver entered Keane's program, and within three weeks she was pregnant. Alexander Malahoff, the intended father, had not even signed his copy of the contract yet.

Under provision 18 of the agreement, Judy swore that she understood the contract and was freely and voluntarily entering into it. But Noel Keane wasn't content to let a surrogate's participation hinge on this clause. He sent Judy to Dr. Parker to ensure that she was capable of giving a voluntary, informed consent. Then, in a somewhat demeaning exercise, he had Judy copy over a paragraph in her own handwriting. The paragraph read: "I understand that the child to be conceived by me is being done so only because the wife of the natural father is unable to bear a child. I, as the natural mother of the child to be born, acknowledge that it is in the best interest of the child for its natural father to have immediate custody. I agree to cooperate fully to place the child with its natural father as soon as possible after its birth." Judy, in her round, clear, sensible handwriting, copied the statement and dated it.

Nine months later, Judy entered Lansing General Hospital and gave birth to a boy who had microcephaly (a small head, indicating possible mental retardation) and was suffering from a life-threatening infection. Dr. Carla Smith asked Judy's permission to treat the baby with antibiotics. Judy disclosed that she was a surrogate mother and that the doctor would have to seek the permission of Alexander Malahoff, the man who had engaged her to bear the child. Judy, complying with the contract, said she felt "no maternal bond."

When Alexander was contacted in New York, he refused to allow the baby to be treated. He wanted a child to hold his marriage together, but now his marriage was even more rocky. The thought of caring for a seriously ill child apparently was just too much.

Who were the legal parents? While in another society this issue might have been resolved by wise elders, in our society it was put to the test of the Nielsens. Alexander, Judy, and Ray (Judy's husband) agreed to argue their case on Phil Donahue's television show.

The Donahue producers had arranged the show for the day that the laboratory would determine, through blood tests, who was the father of Baby Doe. Alexander, Judy, and Ray would get this news for the first time in front of millions of viewers.

The camera panned the trio, none of whom looked eager to pass out cigars. And the loser was . . . Ray Stiver.

Apparently Judy had been told not to have sex with her husband for a month after the insemination but had not been told to abstain immediately before the procedure. She became pregnant by Ray just before the insemination and carried the child, all the while thinking it was Alexander's baby.

The Stivers could have left the baby in foster care or given him up for adoption by strangers. Instead, they pulled themselves together, gave him a name—Christopher Ray Stiver—and took him home to join Mindy, his three-year-old sister.

Within three weeks of the baby's birth, Alexander Malahoff had filed a lawsuit against Judy Stiver, Ray Stiver, and Noel Keane. The well-meaning Stivers had done everything they could throughout the pregnancy to provide the best start in life for the child they thought was Alexander's baby. They had not yet recovered from the shock of unplanned parenthood and the trauma of a child with a serious illness. They weren't sure how they would be able to afford the medications and doctors' visits that young Christopher needed; they couldn't even afford a phone to ensure that they would be able to call an ambulance. And here came Alexander Malahoff suing them for more than $50 million. His claim: since Judy had not gotten pregnant with his sperm, he had been "denied the love, services, affection, and happiness" of the child that Judy had promised to bear for him.

Ray and Judy now had to find their own lawyer to defend themselves. When they did, the legal issues began to snowball. They sued the professionals associated with Noel Keane's program for not advising them to forgo sex before the insemination. They sued Noel Keane for not properly reviewing the procedures used in his program. And, finally, they sued Alexander Malahoff, claiming that his sperm had transmitted the infection that caused Christopher's illness.

The court ultimately sided with the Stivers—saying that if lawyers got into the business of making babies, they had better do it with "heightened diligence."

• • •

I testified in the U.S. Congress, stressing the need for a new law on surrogacy that would ensure proper screening and establish who the legal parents were. Unfortunately, Congress was not ready to allow women to give up their babies for a fee, and the legislators ignored the issue.

Shortly thereafter, however, in Michigan, State Senator Connie Binsfeld created a panel of advisers to draft surrogacy legislation, and she asked me to be on it.

We met in an imposing place: a hundred-room mansion known as Meadow Brook Hall, in Rochester, Michigan, about an hour outside of Detroit. Built in the late 1920s by Matilda Dodge, the widow of auto czar John Dodge, and her second husband, Alfred Wilson, the 80,000-square-foot mansion has been willed to the state for the creation of Oakland University, which uses it as a conference site.

The family the Wilsons had raised in the house was an eclectic mixture of children from John Dodge's first marriage (Matilda was his third wife), children from the union of Matilda and John, and two children whom Alfred and Matilda had adopted.

As the committee members arrived at Meadow Brook Hall, we noticed that the mansion still retained all the touches of a private home: little knickknacks in the rooms, worn carpeting, no locks on the bedroom doors, bathrooms down the hallway. It was an intimate atmosphere, despite the size of the place, in which twenty people could become close and supportive like a family. Or, also like a family, they could have bitter disagreements and jockey for power.

I expected an objective assessment of the issue of surrogate motherhood, but my first few minutes at Meadow Brook Hall should have clued me in to the fact that this would not occur. The antisurrogacy contingent had sumptuous private rooms. The feminists who were more comfortable with surrogacy— Nadine Taub and I—were given a tiny maid's room to share. Worse yet, the room was on the public tour, so every afternoon, a guide would burst into our room with retirees or schoolchildren and point to the stacks of linen that were stored there.

Shortly before our committee began meeting, the Vatican released its "Instruction on Respect for Human Life in Its Origins and on the Dignity of Procreation," criticizing traditional artificial insemination, surrogate motherhood, and in vitro fertilization. Despite all that I had read in law school about the separation of church and state, that document held sway in policy circles. At one Illinois meeting of state lawmakers, I asked when we would get a law about surrogate motherhood. One legislator turned to the Catholic bishop's ethics director in the audience. "When *he* says it's okay," the lawmaker said.

At the Michigan meeting, the religious influence was also unavoidable. At the opening night dinner, when the participants gathered in a circle to begin their first deliberations, State Senator Connie Binsfeld stood up and announced, "Since I didn't order the cherries jubilee for dessert and nevertheless it came for all of us, I think that the flame is the presence of the Holy Ghost and that He is watching over our deliberations."

It was no surprise, then, that Senator Binsfeld opposed any technology that seemed to depart from the traditional nuclear family. Such a requirement, though, overlooked both the blended family experience of the owners of Meadow Brook Hall and the current social reality that most families in the United States do not fit the traditional model. About half of the children born today will spend time in a single-parent household by the time they are eighteen. Yet study after study testifies to their resilience. Children in single-parent homes have cognitive abilities and a level of self-esteem that are at least equal to those of children in two-parent families. The children of surrogate mothers, who for the most part are going into the idealized household of married heterosexuals—the biological father and his wife—could be expected to do at least as well.

The belief that children would be ruined unless they were part of a two-biological-parent family was not borne out in committee members' own lives. Connie's mother had died when she was a very young child, and she had been raised by her father. She was living testimony that people could flourish even if their families were not right out of *Leave It to Beaver*.

The committee took the position that "all human beings have a basic human right to seek to reproduce. The Constitution protects the right against governmental interference, although a compelling state interest might outweigh this right."

Rather than implement that decision, Senator Binsfeld introduced a bill into the Michigan legislature to make paid surrogate motherhood a criminal offense, with the couple and the surrogate facing up to a year in jail and a $10,000 fine.

The lawyers, doctors, and psychologists who fostered surrogacy arrangements could be sent to prison for five years and fined $50,000.

The law sounded like one proposed in the 1960s in Ohio that would have criminalized artificial insemination by donor and subjected all the participants—the doctor, the donor, and the couple—to a fine and imprisonment. That law never passed, and, in fact, at least thirty-five states now have laws to facilitate donor insemination.

The outcome for Binsfeld was much different. Her law banning surrogate motherhood was adopted.

"Under such a law," says Bill Handel, the lawyer at the Center for Surrogate Parenting and Egg Donation in Beverly Hills, California, "a doctor who inseminates a surrogate is as much a criminal as if he held up a Seven-Eleven with a sawed-off shotgun."

When the legislative dust settled, three states had banned surrogacy, thirteen adopted laws prohibiting enforcement of the contracts, and only three offered a modicum of protection to the participants. Only a handful of states clarified who would be the legal parents if there was a conflict. In North Dakota and Utah, the child would go to the surrogate and her husband; in Arkansas, Florida, and Nevada, to the contracting couple.

Despite inadequate regulation, the actual practice of surrogacy continues to thrive, with people engaging in "reproductive tourism," traveling to states with laws most favorable to their desires.

Initially, the procedure involved inseminating the surrogate with sperm from the husband of an infertile woman. The surrogate was both the genetic and gestational mother of the child. More recently, a new form of surrogacy has developed, in which the infertile couple's embryo is implanted into the surrogate mother. The surrogate thus provides only the nurturing from her womb.

In 1991, a forty-two-year-old South Dakota woman, Arlette Schweitzer, became pregnant with her own grandchildren. Her daughter had been born without a uterus. The daughter's eggs and her husband's sperm were used to create embryos, which were implanted in Arlette.

In Italy, in 1997, a surrogate carried embryos from two couples. DNA testing sorted out which baby belonged to which couple. She claimed she was not paid, but imagine the incentives. If a woman can make $10,000 as a surrogate for one couple, she could double her earnings carrying babies for two. Yet the risks to her and to the children are much greater with multiple births.

In the last few years, homosexual couples have also begun to use surrogacy in what is labeled a "gayby" boom. A West Los Angeles company, Growing Generations, is helping gay men achieve their goal of becoming fathers. The three-year-old company, believed to be the only surrogacy agency in the United States catering only to the gay community, has matched at least fifteen male couples with surrogate mothers. One surrogate from Iowa had never met a gay man before. She had dialed up Growing Generations and volunteered after reading *And the Band Played On*, a book about the AIDS epidemic, in a community college course.

To prepare the couples for the potential tough questions their family relationship might raise, the psychologist at Growing Generations asks: "What will you tell your child about the role of the mother? How will you help your child explain to other kids why he has two fathers?"

Growing Generations, www.Surrogacy.com, New Life, Creating Families, Inc., and www.iwhost.net/surrogates are just a few of the surrogate matching services available on the Internet. Surrogacy.com responds to four hundred e-mails a day. As a result of these Internet matches, at least thirty babies have been born through traditional or gestational surrogacy. One reason that couples are surfing the Net to locate a surrogate is to save money by avoiding the finder's fee charged by agencies, which can run between $10,000 and $20,000. Dispensing with the assistance of an agency, however, could be problematic for some consumers. Before matching intended parents with a surrogate, many agencies will arrange for medical and psychological evaluations of the surrogate; they also will make sure that intended parents have been screened, too. Without the guidance and expertise of an agency, parties to a surrogacy arrangement may not realize what type of screening is necessary. For example, the parties may not grasp the importance of screening the surrogate's partner for HIV or other sexually transmitted diseases. Additionally, the participants may not recognize the need for independent legal counsel. Often, the surrogate and intended parents who find one another via the Internet live in different cities, further complicating arrangements for screening and legal advice.

Disputes over parenthood continue to plague the courts since most surrogacy arrangements occur in states without legislative directives about parenthood. In one case, a surrogate named Elvira Jordan was artificially inseminated with the sperm of Robert Moschetta. When the Moschettas began having marital problems, the surrogate insisted the couple have marriage counseling. After the child's birth, the Moschettas split up and Elvira Jordan sought custody. The judge held that the surrogate had parental rights and Cynthia Moschetta did not, even though she had been raising the child for seven months. Under the court's arrangement, the child was given to the surrogate weekdays during the day and to Robert Moschetta in the evenings and on weekends. Cynthia was cut out entirely.

John and Luanne Buzzanca have taken reproductive technology to its logical extreme. Both were plagued by infertility problems, so they went so far as to hire a sperm donor, egg donor, *and* a surrogate mother. The resulting child would have five parents, including themselves, a royal flush of parenting.

But then John Buzzanca filed for divorce before the surrogate gave birth. He also claimed he shouldn't have to support the child-to-be because he had no genetic tie to it. The trial court judge ruled in his favor, suggesting, astonishingly, that since John and Luanne Buzzanca had used an egg donor, sperm donor, and surrogate, the baby, Jaycee, had no legal parents. An appellate court said, "We disagree. Let us get right to the point. Jaycee never would have been born had not Luanne and John both agreed to have a fertilized egg implanted in a surrogate." John and Luanne were the legal parents.

But babies shouldn't have to go through years of litigation (as Jaycee Buzzanca did) to determine who their legal parents are. State legislatures need to provide protections for all the participants in surrogate mother arrangements.

I am reminded of this need most fervently when I receive phone calls like the following: "We advertised in a downstate newspaper for a surrogate mother and invited her to move in with us for the process. She's now here. What do we do next?"

8

The X Vials

❧

Nature smiles on Escondido, California. The turquoise skies yield just the right combination of sun and rain to produce extralarge oranges and bountiful pastures for grazing horses. The temperature cooperates as well, with mild days and a nighttime coolness perfect for sleeping.

In this land of nature at its best, Robert Klark Graham launched a plan to create human beings at their best. In the basement of an old well house on his ten-acre Escondido farm, Graham established the Repository for Germinal Choice. There he collected the sperm of Nobel prize–winning scientists and, later, other geniuses. The destination of these tadpolelike creatures: the wombs of highly intelligent women who wanted supersmart kids.

Graham, a wealthy optometrist and inventor, did not provide this valuable sperm to just any interested or intelligent female; he believed that only one out of every fifty women was qualified to bear the children of his Nobelists. The women had to be members of Mensa, an international society of people who are in the brightest 2 percent of the population, according to certain IQ tests. When a woman was chosen, Graham would send her the precious fluid in a liquid nitrogen tank via overnight mail.

When Graham first opened his clinic in 1979, I was curious about the claims he was making. He advocated a scientific approach to a better world, a perpetual Escondido, of brighter, more creative, happier people. The vision was seductive. No more dullard bosses or store clerks who can't get your order straight. No more money being drained away from creative projects to be spent on remedial education. A chance for smart, charming people (presumably like Graham and myself) to have some say in how society works. As I read about Graham's repository, I wanted to know more about the secrets of parenting he claimed to possess.

Getting through to the man was not easy. Women were calling him or writing to him from all over the world. Scientists who believed they were geniuses were vying for the opportunity to donate sperm and sire American geniuses.

Graham espoused the belief that recent trends in fertility would doom the human race to a downward evolutionary spiral. This was due, in Graham's terms, to the low rate of 1.8 birth per couple of "our more able citizens (European and American)," which is "significantly below the 2.1 rate necessary for them just to maintain their numbers." This "failure" was worse because it did not compete with the 4.1 reproductive rate of "the exploding world population."

Graham believed that civilization would be advanced more quickly if the "more able" reproduced at higher levels to reverse the decline in intelligence (a "chronic affliction of civilization"). He relied on the work of Arthur Jensen, who, according to

Graham, "concluded that variations in intelligence are about 69% hereditary, 25% environmental, and 5% attributable to test error." Therefore, the billions of dollars spent on education, said Graham, were wasted because "you have to start over again with each generation." With selective human breeding, however, "hereditary improvement continues on for generations."

With this philosophy in mind, and millions of dollars burning a hole in his pocket, Graham decided to open his Hermann Muller Repository for Germinal Choice to give babies a "genetically advantaged start in life." Muller had won the Nobel prize in 1946 for his work demonstrating that X rays caused genetic mutations. Muller had proposed such a sperm bank decades earlier, suggesting that it might someday be considered to be unethical *not* to use donors chosen for admired traits. Muller asserted that "foster pregnancy, which is already possible, will become socially acceptable and even socially obligatory." He said that "it will seem wrong to breed children who mirror parents' peculiarities and weaknesses. In the future, children will be produced by the union of ovum and sperm, both derived from persons of proved worth, possibly long deceased, and who exemplify the ideals of the foster-parents." But long before his death in 1967, Muller had severed his relationship with Graham, and his widow, Thea, was scandalized when Graham named the bank after her husband. She said her husband had withdrawn his support for the project before his death because of Graham's insistence that they attempt to create only intelligent children, rather than altruistic ones.

Frustrated after leaving a variety of messages to try to get Graham to call me back, I decided to pose as a woman who was interested in his services. I took the Mensa test, became a member, and left this message: "I'm a well-adjusted, childless Mensa woman and Yale-educated lawyer." Soon he returned my call.

Within a few days, in the spring of 1979, I was sitting in his Escondido office. I eavesdropped while he dealt with a call from a potential donor from Europe. "I haven't even a Nobel prize yet, but I'm the smartest man in France," the caller said to Graham.

The donors filled out a nineteen-page application form, with questions such as "How often do you lose your temper?" and "Have you ever had delusions of greatness or omnipotence?" To collect the sperm, Paul Smith, Graham's associate, visited the scientists in their offices.

While Graham pondered whether he should anoint the Frenchman to fatherhood, I looked through the materials he had given me, a potential "applicant." The first was a letter that began: "The Repository for Germinal Choice brings to qualifying women a new resource: the freedom to choose the father of their child or children from among the most creative scientists of our time." The application form asked me questions about my IQ, achievements, and medical background. I would be screened by Graham and two San Diego physicians. I was not informed why these particular doctors were qualified to pass judgment on my ability to mother.

Once I had chosen a donor, I would be asked to sign a witnessed agreement to "respond fully to questionnaires regarding offspring which the Repository may submit from time to time." Then a two-inch vial of sperm, frozen in liquid nitrogen, would be sent to me in a special two-foot-high container. The shipment would be timed to arrive at the point when I ovulated.

As Graham sifted through the dozens of new applications that had arrived that day, I looked at the sheets of information listing the ostensibly inheritable characteristics, but not the names, of various genius donors. On the bottom of these forms were Graham's own comments (such as "a very famous scientist—a mover and a shaker—almost a superman").

Finally Graham turned his attention back to me. "I'm carrying out the ideas of the late Nobel prize winner Hermann Muller, who advocated a sperm bank of donations from men of exceptional achievements," Graham told me. "But I've limited it to Nobelists in science because I don't want people to say, 'Who are you to play God?' This way I don't; I leave it up to the Nobel Prize Committee."

Yet Nobel laureates were skeptical of Graham's plan. He tried to solicit donations from Nobel laureate Dr. Max Perutz of Cambridge, England. "I wrote back saying I was small, bald, and plagued with back trouble," said Perutz.

I asked if I could visit the sperm bank itself. On the drive from the office to his Escondido farm, Graham explained that the facility had the potential of creating 100,000 children. We passed a wild-animal compound. "They're doing the same thing here that we're doing at the repository," Graham explained as he drove. "Breeding to keep certain animals from becoming extinct. They're even saving the cells of the rare ones so that when cloning is perfected, they can clone them."

Graham ran through his own intellectual pedigree as we began exiting the highway. He was a Mensa member who had applied his smarts, he said, to invent shatterproof eyeglasses. Now a multimillionaire, he employed twenty-five staff members who did nothing but handle his investments, from oil wells to apartment buildings. He said he had done his part to upgrade the species by having eight children. As he told me about his intelligence, he began exiting the expressway by doing a U-turn onto the on ramp. Cars swerved and honked, but Graham didn't appear to notice.

My heart was still racing as we pulled into a gas station.

Graham approached a pump, but instead of filling the gas tank, he took a liquid nitrogen tank out of the backseat and filled it. This was the substance that was used to preserve all the sperm. But Graham misdirected the nozzle, and the smokelike liquid nitrogen started squirting all over the pavement.

At the farm, Graham carried the liquid nitrogen to the well house. I descended the stairs beside him and entered what looked like a suburban rec room. Later, I asked California lawyer Bill Handel if there were any public health regulations stipulating how sperm must be stored. "No," he replied. "If I wanted to open a sperm bank in the deli next to the pastrami, there would be nothing to stop me."

The lack of regulation was underscored when Graham's employee Paul Smith split off to form his own genius sperm bank, Heredity Choice, in 1984. Thirteen years later, in an extraordinary move, California public health officials shut it down. There were three reasons for the action: no running water, lack of basic hygiene, and the charge that Smith, who breeds dogs as well as humans, "commingles dog semen samples with human samples." Smith is challenging the state's action to close Heredity Choice, arguing that "none of my clients has ever had puppies from sperm I have supplied."

"Intelligence is the only thing that sets man apart from the beast," Graham told me after we emerged from his sperm bank and walked back across the lawn, past the eucalyptus trees and peach blossoms. "For many years, man evolved through the survival of the fittest so that only the most intelligent lived to bear children, but now technology lets those with genetic defects live and reproduce. And while bright women are postponing having children or not having children at all, we're financing poor women to have them."

Graham stopped for a moment to pluck an extralarge

tangerine from one of his trees. He gave me the opportunity to do the same. I chose one, but obviously he felt that I had not chosen well. He took away the tangerine I had picked and gave me a rounder, more perfect specimen instead.

• • •

In the late 1800s, a majority of the geneticists in this country believed that one could extend genetic principles to explain human behavior. Traits such as feeblemindedness, criminality, pauperism, prostitution, and intelligence were thought to be single gene defects. People began to make private choices about whom to marry on genetic grounds, to avoid having a child with a disfavored trait. In 1910, the Eugenic Records Office was established in Cold Spring Harbor, New York, which trained field-workers to collect family histories from people around the country. By 1924, data on people had been entered on around three quarters of a million cards, and people made inquiries to the office about whether particular proposed marriages would be eugenically appropriate.

Genetic theories quickly served as the basis for proposals for social and legal reform. The prime thrust of the reforms was to prevent people with presumably undesirable genes from reproducing. The chairman of the Department of Psychology at Harvard University advocated "the replacement of democracy by a caste system based upon biological capacity with legal restrictions upon breeding by the lower castes and upon inter-marriage between the castes." Federal and state legislatures took the teaching of genetics to heart. They passed laws to prevent people with presumably undesirable genes from reproducing, on the grounds that the care of the unfit (such as the mentally disabled) was draining society's resources.

Between 1907 and 1931, thirty states in this country tried to legislate "worthiness" for childbearing by passing laws to sterilize people who were considered to be "feebleminded" or to have criminal tendencies. The operation became known as a "Mississippi appendectomy" because it was used most frequently on poor people from the South. By 1964, over 64,000 people received these "appendectomies," sometimes against their will and often without their knowledge. Since then, the idea that criminality is "inheritable" has been discredited. And as for the so-called feebleminded who lost their ability to bear children, many of them would be considered normal today.

The eugenics movement also had been aimed disproportionately against women, who were more often institutionalized and sterilized than men. In *Buck v. Bell*, U.S. Supreme Court Justice Oliver Wendell Holmes, otherwise a champion of individual rights, authorized the sterilization of a woman, Carrie Buck, on the grounds that she was feebleminded. Holmes stated: "We have seen more than once that the public welfare may call upon the best citizens for their lives. It would be strange if it could not call upon those who already sap the strength of the State for these lesser sacrifices . . . in order to prevent our being swamped with incompetence. It is better for all the world if, instead of waiting to execute degenerate offspring for crime, or to let them starve for their imbecility, society can prevent those who are manifestly unfit from continuing their kind. . . . Three generations of imbeciles is enough."

In a stunning research project, historian and lawyer Paul Lombardo showed in 1985 that Carrie Buck was not in fact an imbecile. She had done well in school, as did her daughter. Rather than being institutionalized because she was feebleminded, she had been institutionalized because she was considered to be "immoral" for having a child out of wedlock. Yet that pregnancy was the result of being raped by the nephew of the foster parents with whom she lived—the very people who committed her to the institution! The doctor who sterilized her, notes Lombardo, was "obsessed with placing checks on sexuality and propagation," and Buck received appallingly poor legal representation. Her lawyer, who had been part of the institution's board that authorized her sterilization, did not call any witnesses or introduce any facts to challenge the characterization of his client as feebleminded.

Graham explained to me that *his* plan was different, since it attempted to increase the number of people at the top of the IQ scale rather than prohibit parenthood among those who are not as smart.

"What happened in Nazi Germany couldn't happen here," he said. Rather than using violence, he advocated something society already seems to accept—social and economic incentives.

In addition to offering mothers the opportunity to choose high-status biological fathers for their children by use of a sperm bank, Graham encouraged rewarding "worthwhile" couples for having kids. For instance, Graham suggested "living remembrances," where a person leaves money in his will to a foundation (such as Graham's Foundation for Advancement of Man), so that the foundation could choose a "superior" young couple to be given the money to defray the cost of having a child. The couple would then name the baby after the person giving the bequest.

Graham also suggested that the government encourage higher-IQ couples to have many children by building special suburban housing developments of attractive three-bedroom homes. Rent for such units would be nominal, but the *only* people who could be allowed to live there would be couples who pass a certain test of superiority and agree to have a child at least every two years.

Stanford University professor William B. Shockley, who won a Nobel prize in physics in 1956, agreed with the idea of economic incentives to clean up our gene pool. Shockley proposed that our society should think about paying people he considered genetically inferior if they agreed *not* to have children. The sterilization payment rate: $1,000 for each IQ point below 100.

At last, Graham focused on my candidacy. His application limited his services to married women—and I was still single when I first visited him. "I'd be happy to consider you as a candidate," Graham said to me. "I need your legal skills, though. Come up with a rule where I can give sperm to you, but not have to give it to an unmarried black woman."

• • •

Given his reverence for scientists, Graham was not going about his plan in a particularly scientific way. For instance, Graham didn't ask about the IQ or achievements of the Nobel laureates' existing children or their ability to function in society; he let Nobelists donate as long as their previous children were physically healthy. As for the passing on of a "science gene," I learned that Nobel prizes were more likely to run in laboratories than in families. Shockley had managed to insult both his wife and children when he told *Playboy* that his own offspring were "a regrettable regression to the mean."

The pseudoscientific nature of Graham's endeavors made many Nobelists balk at providing sperm, and Graham began to seek donations elsewhere. Later, Graham rationalized that the Prize winners' sperm was old, anyway, and therefore would have been poor for fertility. Graham started advertising for donors in the Mensa magazine, with the help of a Mensa eugenics group. Ten years after the sperm bank opened, there were no Nobel sperm donations, and the Repository was relying on the donations of young, healthy, intelligent (white) men.

One of the most popular fathers is a handsome Olympic athlete who, according to Graham, "has the most superb neuromusculature almost in the world. He is worthy of multiplying."

Back in Chicago, my research team and I tried to figure out what sport a smart guy could have conquered.

"Luge," said Michelle.

"No, Ping-Pong," said Nanette.

One of Graham's prize catches, in the mid-1980s, was a Harvard-trained physicist—Richard Seed—who pioneered artificial embryonation and later became known for launching a crazy attempt to begin human cloning. Probably Seed's macabre 1998 television appearance, in which he threatened in a Draculalike drawl to steal some of Ted Koppel's blood and clone him against his will, led some of the women who used Graham's bank to worry about the parentage of their offspring.

"I am going to sneak down there, accidentally jab a needle into your arm, get some of your blood, and then I can clone you without your permission," Seed told Ted Koppel on *Nightline*, where I was the other guest.

"Well, that would make for an interesting lawsuit," Koppel replied.

• • •

Bernard S. Strauss, professor of molecular genetics and cell biology at the University of Chicago, explained to me that

Graham's plan had a very simplistic view of human biological endowments. "Intelligence cannot be traced to a single gene. The hereditary input comes from a variety of chromosomal signals. The children of Nobel prize winners will vary in intelligence, even in those cases when the Nobel prize winners' wives are also of exceptional intelligence."

"One of the problems with using sperm from Nobel prize winners," added Reuben Matalon, a pediatric geneticist at the University of Texas Medical Branch in Galveston, Texas, "is that most of them are older men. Until recently, birth defects were traced to the old eggs of mothers over forty. But research is showing that such chromosomal abnormalities may also be linked to older fathers, especially those over age sixty. Even if the Nobel prize winners had healthy and intelligent children with their wives when they were young, there is no guarantee that the children created today by artificial insemination would also be healthy or intelligent."

The real tragedy of a eugenics plan is that it may in many cases overlook the very people who would be the best parents. A woman who has sensitivity, love, and commitment might be discouraged—or even prohibited—from becoming a mother merely because of some genetic "flaw" such as asthma or diabetes or a poor mark on a particular IQ test.

When I asked Graham if he made any suggestions to women who use his sperm bank about how to raise their children, he told me, "That would be presumptuous. I'm not an expert on that. The only thing I am an expert on is eyeglasses."

Graham is not the only purveyor of gametes who is working to upgrade the human species. "Everybody wants as much information as they can get to make a wise decision, whether it's purchasing a home, or buying a car, or going on a vacation, or selecting a sperm donor," says David Towles, marketing director of the Xytex Corporation sperm bank.

"Why is it okay for people to choose the best house, the best schools, the best surgeon, the best car, but not try to have the best baby possible?" the parents of a child conceived with high-IQ sperm asked a *Sacramento Bee* reporter.

"You look, and you eliminate things that just aren't interesting to you, such as, one of the profiles had on it that they had a Richard Nixon nose," said Jacqueline Teepen, who appeared on *Good Morning America* with me to discuss her use of smart sperm. "That wasn't an interest of ours. We wanted somebody with hazel or blue eyes, we wanted the bachelor's degree to be finished, working into the master's or even a Ph.D. program.

"I think it's wonderful," she continued. "I think the ability to select characteristics is simply wonderful."

Columbia-Presbyterian Medical Center in Manhattan will customize an embryo by choosing both an egg donor and sperm donor to match the desires of the parents. The clinic, run by Mark Sauer, allows couples to "adopt" an embryo for $2,725. The embryos are made from the surplus of donated eggs from women who have undergone fertility treatment for themselves, and from sperm donated to the clinic.

I have also encountered couples who wanted to create a blueprint for their brave new baby by a surrogate mother with particular traits. One couple, for example, wanted a surrogate who was a cellist. They wouldn't accept a surrogate who played any other musical instrument. A single man contracting with a surrogate wanted one who was a cross between Eleanor Roosevelt and Brigitte Bardot. Amazingly, attorney Noel Keane found a surrogate who matched that description. The deal never went forward, though. The woman was too headstrong to agree to the terms that were offered in the contract.

Celebrity sperm and egg donors are also highly appealing. When one researcher jokingly offered to sell Mick Jagger's sperm, women wrote to him with their orders from all over the world.

In the years since I visited Graham, I have kept track of his operation. More than two hundred children have been born using Graham's sperm collection. They live throughout the United States and in other nations—England, Germany, Canada, Italy, Egypt, and Lebanon. Most of their identities have been kept secret. Other than Graham, only one individual knows who they are. In 1983, *California Magazine* published a photo of that person: the Federal Express driver serving Escondido, who has seen the name and address of every recipient of liquid nitrogen tanks sent by overnight delivery from Graham's bank.

Yet I worry about the children who were made to order with genius sperm. Would everyone be waiting for $E = mc^2$ to come out of their mouths?

Victoria Kowalski was the first child born using Graham's sperm bank. After her birth in April 1982 to Joyce and Jack Kowalski of Scottsdale, Arizona, her parents sold the story rights to the *National Enquirer* for $20,000. "The odds are very good that our little girl will turn out to be a genius," Mrs. Kowalski told the *Enquirer*. "I imagine her as a child studying college textbooks."

The news of this bundle of joy was received with horror by Joyce's two children from a previous marriage, Donna and Eric, who were being raised by their father. Joyce had lost custody of those children after she and her new husband, Jack, had abused them—in an effort to make them smarter. Young Donna had been forced by her mother and stepfather to wear a sign that said "Dummy" on her forehead. Donna and Eric were given extra homework by the Kowalskis—and then whipped with a belt or strap when they made a mistake. "They said they were trying to help us," said Eric. "They wanted us to work and be smart."

It was chilling, though, to think about the life ahead for Victoria—certainly in light of the expectations her parents have for her. "We'll begin training Victoria on computers when she's three, and we'll teach her words and numbers before she can walk," Jack Kowalski told the *Enquirer*.

I was also troubled by the couple who appeared on *Good Morning America*. The mother pointed to the infant rolling around on a bed and described how intelligent the baby was. Yet the child looked quite ordinary to me, and I wondered what would happen if the child was not on the far end of the bell curve.

Doron Blake, who was born August 24, 1982, is the star in the Repository stable. With an IQ of 180 and a full scholarship to Phillips Exeter Academy, Doron is, in theory, what Graham intended to create when he offered "genius" sperm to women. He is reported to have used a computer at age two, read Shakespeare at age five, and composed concertos after finishing his homework at age thirteen. However, BBC reporters who filmed Doron at age three found that "he showed no aptitude for any musical instrument, nor was he very prodigious on a home computer which he thumped happily."

Doron knows about his origins. "I've never felt that I'm part of an experiment or master race," he says. "My mom wanted a child because she loved children. She just wanted to make the best choices when it came to having me."

As his mother, Dr. Afton Blake, puts it: "The fact that the man was healthy appealed to me, of course. . . . Why not have a child born with an advantage in an increasingly difficult world?" In fact, Doron thinks it is "cool to be different." He does, however, wonder what his sperm bank father thinks of him: "What role do I play in his life? You know, what am I? Am I your son, or am I your, you know, sperm bank offspring?"

9

Admission Standards for Birth

❧

Three days after Alexis Ferrell was born, blood tests indicated that she might have Fanconi's anemia, a life-threatening disease. But her doctor, Dr. Kenneth Rosenbaum, never looked at the test results. Nor did he look at two subsequent blood tests—both of which were marked on laboratory reports to indicate an abnormality.

When Alexis was three, she was hospitalized with pneumonia. A new doctor diagnosed her immediately, confirming Fanconi anemia, and indicating that without a bone marrow transplant she wouldn't survive into adulthood. Her mother sued Rosenbaum.

This is where it gets tricky. At the time of Alexis's birth, there wasn't any existing treatment Dr. Rosenbaum could have given that would have made her better. His lawyer argued that since his mistake had no effect on the outcome, the doctor had caused no harm.

Alexis's mom, Susan Ferrell, disagreed. She made the following claim: If the doctor had diagnosed her daughter at birth,

Susan wouldn't have seperated from Alexis's father. Instead, she would have conceived another child with him to serve as a matched bone marrow donor for Alexis.

The defense wondered why she didn't just call the father and ask for his help. But Alexis's dad was now homeless, on the other side of the country, and couldn't be reached. Susan Ferrell argued that she'd been deprived of her only hope—having a child with him before he dropped from sight.

The defense argued that this possibility was just too remote. What if he hadn't wanted to stay together? What if he refused to have another child with her?

But the judges on the D.C. Court of Appeals were no strangers to reproductive technologies. They pointed out that Mr. Ferrell wouldn't have to reconcile with the mother. He could have merely donated sperm to her.

"The proper diagnosis," wrote the court in 1997, "would then have allowed the Ferrells to produce a natural donor sibling who would now be life-saving."

Over one hundred children have been conceived as prospective donors. Couples get pregnant, have the fetus tested prenatally to see if it is a match, and then abort and try again if the new child cannot serve as a donor. There are no figures on how many fetuses have been aborted under these circumstances. One couple who chose not to abort created three additional children, yet none was a suitable match. In another family, a baby was created to be a donor—but the infant turned out to have the same rare metabolic disease as the existing child.

New reproductive arrangements create new tensions for families. When a remarried woman learned that a child from her first marriage needed a bone marrow transplant, she used artificial insemination with sperm from her first husband to conceive a child. But how will her second husband feel about her having a child with the first? Who will raise the new baby? For doctors pushing conception as a medical treatment, these questions are considered outside their purvue.

In some cases, an existing family member is a match for the ill child but does not want to go through the painful procedure of bone marrow extraction. Does it seem fair to create a child and force him or her to do something that an existing sibling, cousin, or parent won't do? And where should we draw the line? Should parents conceive a second child to be a kidney donor for the first?

In many parts of the world, technology is making the admission standards for birth tougher and tougher. In India, China, Taiwan, and Bangladesh, technicians with portable ultrasound machines go from village to village scanning pregnant women who are desperate to learn whether they are carrying a boy. Many abort when they fail to see a penis on the tiny out-of-focus screen. In Bombay alone, 258 clinics offered amniocentesis for sex selection. In one study of 8,000 abortions in India, 7,999 were female fetuses, leading human rights activists to protest this clear evidence of "gyne"cide. In China, when the one-child policy was strictly enforced, families so preferred males that the sex ratio changed to 153 males for each 100 females.

At Dr. John Stephens's clinics in California, Washington, and New York, Western couples too can have prenatal testing for sex selection. One Australian client actually terminated a pregnancy because she couldn't get to Stephens's clinic in time to learn the sex of the fetus. She carried the next pregnancy to term when Stephens vetted it as a boy. Although most couples want a boy, an Israeli couple went to great lengths to have a girl out of fear of losing a son in a military engagement. They aborted a fetus when they learned it was a boy. In the next pregnancy, the wife was carrying twins, a boy and a girl. She used selective reduction to abort only the boy.

Thirty-four percent of U.S. geneticists said they would perform prenatal diagnosis for a family who want a son, and another 28 percent said they would refer the couple to another doctor who would perform such testing. Dorothy Wertz, the social scientist at the Shriver Center for Mental Retardation in Waltham, Massachusetts, who conducted the study, said the percentage of practitioners willing to respond to sex selection request had increased 10 percent from 1985 to 1995. "Autonomy just runs rampant over any other ethical principle in this country," Wertz says. "And it's only going to increase."

At the Genetics and IVF Institute in Fairfax, Virginia, a machine measuring DNA content is used to separate Y-bearing sperm (which produces males) from X-bearing sperm (which produces females, and which has 2.8 percent more genetic material). Sperm of the favored sex can then be used to create a child. The cost is $2,500, less than the $10,000 it costs to use IVF and preimplantation screening for sex selection.

The clinic has published only those results from families who wanted girls—perhaps to make the procedure seem more politically acceptable. It did not disclose how many couples wanted boys. The technique is more accurate for choosing girls (93 percent) than boys (65 percent).

Paul Rainsbury, director of the IVF program in Essex, wants England to change its laws so he won't have to send couples to the Virginia clinic. "We are doing surrogacy and allowing artificial insemination of lesbians. These are far bigger ethical minefields than sex selection," says Rainsbury.

What if a sexual imbalance occurred in the United States, as is now happening in China and India? Sociologist Amitai Etzoni speculated that since women consume more culture and men commit more crimes, sex selection would create a more frontierlike society—with less art and more violence. Since men are more likely to vote Republican, politically there would be a shift to the right.

The overwhelming tilt toward boys is not as pronounced yet in the United States as it is in other countries, but social psychologist Roberta Steinbacher of Cleveland State University worries about the effect on society if couples were able to predetermine their baby's sex. Twenty-five percent of people say they would use a sex selection technique, with 81 percent of the women and 94 percent of the men desiring to ensure their firstborn would be a boy. Since other research reveals firstborns are more successful in their education, income, and achievements than latterborns, Steinbacher worries that "second class citizenship of women would be institutionalized by determining that the firstborn would be a boy."

• • •

Aborting a fetus who is the "wrong" sex and searching for Y-bearing sperm are only the crude first steps in the evolution of designer babies. Soon parents will have more precise tools for genetic engineering.

In vitro fertilization made preimplantation genetic testing possible. The ability to do chromosomal or genetic testing on an embryo led to a further possibility: the chance to treat the embryo when it suffers from a defect or to manipulate its characteristics if by some criterion they don't "measure up."

"It is by bringing the embryo out of the womb and into the light of day that IVF (*in vitro* fertilization) provides access to the genetic material within," Princeton biologist Lee Silver says. "In a very literal sense, IVF allows us to hold the future of our species in our own hands."

The market for genetic enhancements of children is huge—a much more lucrative area for biotech companies to invest in than the treatment of rare genetic disorders. According to a March of Dimes survey, 43 percent of parents would use gene therapy to give their child enhanced physical abilities, and 42 percent of parents would upgrade their child's intelligence level. With over 4 million births in the United States per year, that's a market for genetic enhancement almost as large as that for Prozac or Viagra.

There are approximately 6,000 to 15,000 children in the United States with stunted growth because their pituitary does not produce enough hGH (human growth hormone). Those children are the proper candidates for treatment with a genetically engineered version of hGH. But 20,000 to 25,000 children per year are injected with hGH. Physicians admit that 42 percent of their patients do not have classic growth hormone deficiency; some are just a few inches shorter than average. In a medical research survey, 5 percent of suburban Chicago tenth-grade boys said they had used the hormone.

Human growth hormone has become the forty-third-largest-selling drug in the United States, with nearly a half billion dollars a year paid to the companies that market it, Genentech and Eli Lilly. Even the National Institutes of Health has sponsored a twelve-year study, due to end in the year 2000, giving human growth hormone to normal, healthy children. Not surprisingly, Eli Lilly has given significant funding to NIH to undertake the study, but it is still costing U.S. taxpayers about $200,000 per child.

The hormone doesn't always work, though. Some kids

exceed their expected height by as much as five inches, but others turn out to be shorter than they would have been expected to be without treatment. And according to the American Academy of Pediatrics, there are potential side effects: leukemia, exaggeration of scoliosis, swelling, allergy, and impaired glucose tolerance.

In Charlotte, North Carolina, children were measured at school, and letters were sent to parents suggesting that short kids might need medical attention. Unbeknownst to parents, the nurse running the program was paid $108,000 in consulting fees by Genentech, and her husband was the sole pediatric endocrinologist in the area. One set of parents says the endocrinologist tried to pressure them to try hGH on their eleven-year-old son, who was projected to grow only to five-foot-six. The doctor played on the mother's guilt, asking what she would tell her son when he grew up and learned that he could have been five-foot-ten.

Should society allow doctors to prescribe a potentially dangerous intervention for someone merely because of social prejudice?

"Genetic enhancement is going to happen," says W. French Anderson, the nation's leading gene therapy researcher. "Congress is not going to pass a law keeping you from curing baldness."

Various researchers—including Anderson—have patented gene insertion techniques that would let parents insert desirable genes into embryos to create "better" children. University of Chicago physician-philosopher Dr. Leon Kass has speculated that "the new technologies for human engineering may well be the 'transition to a wholly new path of evolution.' They may therefore mark the end of human life as we and all other humans know it." He points out how jaded we've become, quoting Raskolnikov, the protagonist in Fyodor Dostoyevsky's *Crime and Punishment*: "Man gets used to everything—the beast."

Chillingly, one geneticist told me how genetic engineering could be used to cure racism: "We would just make everyone the same race."

Like the parents of Garrison Keillor's fictional Lake Wobegon, most parents want all their children to be "above average." What happens to our definition of *normal* after genetic enhancement hits the scene?

"There will be many wealthy people willing and eager to pay the price of making their child taller and more beautiful," says Michael S. Langan, a vice president of the National Organization for Rare Disorders. "Eventually there will be discrimination against those who look 'different' because their genes were not altered. The absence of ethical restraints means crooked noses and teeth, acne, or baldness, will become the mark of Cain a century from now."

Some researchers actually propose putting into people genes from other species, such as a gene to photosynthesize. Scientists have put firefly genes in tobacco plants, causing them to glow in the dark, and human cancer genes in mice, leading to the patented DuPont oncomouse. Law review articles have begun to ask how many human genes were needed before a creature would be a protected person under the Constitution.

I asked my law students what they thought.

"If it walks like a man, quacks like a man, and photosynthesizes like a man, it's a man," said a doctor-turned-lawyer in the class. The very boundaries of what is considered human are being challenged by the technology.

The same is true with cloning. After Jonathan Slack cloned headless frogs, other researchers suggested cloning headless humans to serve as organ donors. They argued that, with no brain, such creatures would not be considered persons under the law.

• • •

Philosopher Peter Singer and Australian lawmaker Deanne Wells suggest limiting the power of parents to choose their children's traits. They advocate creation of a governmental body to consider parents' proposals for genetic engineering. The body would consider whether the proposed form of genetic engineering would, if widespread, have harmful effects on individuals and society. Part of Singer and Well's concern is that "if there is pressure on individuals to compete for status and material rewards, the qualities that give children a winning edge in this competition are not necessarily going to be the most socially desirable." The gene for greed might help an individual get ahead on Wall Street, but that might not be the best for the rest of us.

The loss to individuals and society if genetic tampering becomes routine is hardly ever mentioned. Gene therapy to eliminate sickle cell disease is being investigated, even though carrying a single copy of the sickle cell gene is beneficial—it protects against malaria.

At James Watson's Cold Spring Harbor Laboratory, I was part of a conference addressing the provocative question, What if there was a gene that was bad for the individual but good for society? The gene at issue was for manic-depression, a disorder from which many artists and writers have suffered. Some participants at the meeting suggested society would lose out if all manic-depressives were eliminated before birth.

Biologist Lee Silver was not convinced by such an argument. "If the manic-depressive Edgar Allan Poe were never born, we wouldn't miss 'The Raven.' Likewise, we don't miss all of the additional piano concertos that Mozart would have composed if he hadn't died at the age of thirty-four."

The legal and bioethics communities are coming up empty even trying to think of analogies to guide them in this area. Is it "cheating"—akin to taking steroids in sports—for parents to give their kids the genes for height? Or is it more like buying children computers, sending them to soccer camp, or giving them music lessons?

There are medical dangers from gene manipulation, which should give pause to parents who are considering using the technology to upgrade their otherwise healthy kids. When new genetic material is inserted into a cell, it can damage an existing gene, causing it to fail to function. Five percent of mice given gene therapy have such harmful mutations. Should parents and doctors be allowed to take that risk with their children-to-be? And who would "own" the genetically manipulated child?

Researchers from Baylor College of Medicine in Houston have applied to the European patent office for a patent on women who have been genetically engineered to produce pharmaceutical products in their breast milk. So far, the researchers have only used the technique on animals, but they included women in the application, says their lawyer, because of the possibility that whole people may someday be patentable.

Paul Braendli, head of the European patent office, responded with a terse statement: "Human beings are not patentable." But

Baylor is appealing, saying human patenting is not explicitly prohibited under European law. Braendli replied that according to the European Patent Convention, patents must not contravene "public order" or "morality."

Interestingly, under U.S. law, there is no such exception. "Anything under the sun that is made by man" is patentable, according to the U.S. Supreme Court, no matter what the impact on society.

• • •

When Shauna Curlender, a seemingly healthy child in California, degenerated into seizures, blindness, and neurological decay by age three, her parents sued their doctor and the genetics lab that erroneously told them they were not carriers of the gene for Tay-Sachs disease. Their claim: if they had known Shauna would have this painful disease, which would kill her before she reached kindergarten, they would have terminated the pregnancy before she was born.

They won their case, as have parents in twenty-one other states when they said they would have aborted the fetus if they had received correct genetic information. In Alabama, the *sister* of a child with a genetic disorder was able to sue the doctor for the damages that came from "diminished parenting" due to the attention her folks gave to her sick sibling.

But the Curlender case added an even more unusual twist. Shauna herself also sued the doctor and lab claiming "wrongful life." Her attorney argued that Shauna would rather not have been born, rather than born with a painful, fatal disease.

Similar claims in other states had been soundly rejected. "Whether it is better never to have been born at all than to have been born with even gross deformities is a mystery more properly left to the philosophers and theologians," said a New York appellate court. "Examples of famous persons who have had great achievement despite physical defects come readily to mind, and many of us can think of examples close to home," echoed the New Jersey Supreme Court. "A child need not be perfect to have a worthwhile life."

But the California Supreme Court ruled that Shauna had a valid claim. The judges pointed out that, in recent years, doctors had gained the knowledge to avoid, in the court's words, "genetic disasters" and decrease the health care system's burden of caring for children with genetic disease. Since then, New Jersey and Washington have followed suit and also ruled in favor of children's wrongful life claims.

Shauna Curlender's right to sue seemed so clear that the California Supreme Court raised a possibility that none of the lawyers had even requested. The court said that if a seriously ill child's parents had not tested her before birth and aborted her, she could also sue her *parents* for wrongful life.

"If a case arose where, despite due care by medical professionals in transmitting the necessary warnings, parents made a conscious choice to proceed with a pregnancy, . . . we see no sound public policy which would protect those parents from being answerable for the pain, suffering, and misery which they have wrought upon their offspring," wrote the court.

The California legislature was appalled by the possibility of kids suing parents for giving birth. It immediately passed a law banning such suits, as did eight other states. And an Illinois court turned down a child who tried to sue his father for getting his mother pregnant out of wedlock, leading to his being born illegitimate. The court said it would open up the floodgates for suits by children who were dissatisfied because they were born the wrong color, the wrong race, not sufficiently healthy—or from a parent with an "unsavory reputation."

• • •

At the Genetics and IVF Institute in Fairfax, Virginia, director Dr. Joseph Schulman offers even newer genetic testing techniques. One of his staff doctors removes a single cell of a couple's eight-cell embryo, then freezes the remaining seven. With genetic testing techniques, he will check for Down syndrome, cystic fibrosis, and a host of other genetic conditions. If the embryo gets a clean bill of health, it will be defrosted and implanted into the mother-to-be.

Long before Congress's decision to fund the massive Human Genome Project, Schulman was the first director of the Medical Genetics Program at the National Institutes of Health, remaining on the faculty there from 1973 to 1983. Described even by his critics as a "hard-driving, ambitious genius," Schulman was the first American to travel to England to study with IVF pioneer Robert Edwards—four years before the birth of Louise Brown. When Schulman looked at a human embryo under a microscope in 1974, he was one of only a handful of researchers in the world who had seen the beginning of human life.

Schulman returned to the United States to set up an in vitro fertilization program at the National Institutes of Health, but the Nixon administration put a ban on government funding for any research involving human embryos. The pro-life lobby was vehemently against it. Schulman was told at first the moratorium was temporary, but, in fact, to this day, NIH refuses to support IVF research.

Frustrated by the fact that in vitro fertilization and genetics research were being held hostage to the abortion debate, Schulman set out on his own. In 1984, he formed one of the first privately owned, for-profit infertility clinics: the Genetics and IVF Institute. Now his institute performs hundreds of IVF procedures a year for women up to age fifty-five and processes up to 7,000 samples of amniotic fluid for prenatal diagnosis of more than two hundred inherited conditions.

With more than two hundred employees and franchises in Ohio and Florida, Schulman operates not so much a clinic as an industry. The latest technique added to his menu of services, preimplantation genetic screening, combines his expertise in genetics with his background in IVF.

Prenatal genetic testing has become the newest consumer good, with competitors to Schulman popping up like Starbucks franchises. Parents are being given unprecedented power over controlling the genetic makeup of their children, and the exercise of that power is now routine.

Some doctors are even calling patients out of the blue to promote their services. Dr. Gary Hodgen of the Jones Institute called Renee and David Abshire, whose child had died of Tay-Sachs. He offered to perform preimplantation screening on their embryos so they could avoid having a second child with the same disease.

"Strangers in the supermarket, even characters in TV sitcoms, readily ask a woman with a pregnant belly, 'Did you get your amnio?'" notes disability rights activist Martha Saxton. Even a government agency, the Office of Technology Assessment of the U.S. Congress, endorsed this approach. After describing new genetic tests, an OTA report concluded that "individuals have a paramount right to be born with a normal, adequate hereditary endowment."

The vehemence with which many Americans hold that belief

came to light in the reaction to the pregnancy of California anchorwoman Bree Walker. Walker is affected with ectrodactyly, a mild genetic condition that fused the bones in her hand. When she decided to continue her pregnancy of a fetus with the same condition, a radio talk show host and her audience attacked the decision as irresponsible and immoral.

"Prenatal screening seems to give women more power," says disability rights activist Laura Hershey, "but is it actually asking women to ratify social prejudice through their reproductive 'choice'?"

In a U.S. survey, 12 percent of potential parents said they would abort a fetus with a genetic predisposition to be fat.

10

Island Secrets

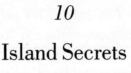

T he island of Sardinia was settled in the Neolithic era by
people who found that obsidian (black glass from the vol-
canic ashes of mountains like Mount Arci) was useful for
making tools. They mated with the Nuraghesi, who are
believed to have sailed there from the eastern Mediterranean
around 3000 B.C. The island was successively invaded and occu-
pied by Greeks, Romans, and Phoenicians, and later looted by
the Vandals, Byzantines, and Arabs. In the Middle Ages, first
the maritime republics of Pisa and Genoa tried to win Sardinia,
then the Spanish, leading to the mixture of architectural styles
and language.

The old portion of the island has Roman ruins, Pisan watch-
towers, and Spanish homes with wrought-iron balconies. In its
National Archaeological Museum, there are hundreds of bronze
miniatures showing Nuraghesi life 2,500 years ago. But the
Nuraghesi had a greater impact on today's Sardinia than even
these breathtaking miniatures would indicate. They brought
with them the beta-thalassemia gene, which causes a deadly
childhood anemia.

"You can tell by the skeletons of the Nuraghesi that the disease is very old," says Dr. Antonio Cao. I had come to visit his Instituto di Clinica e Biologia dell'Età to learn more about the Sardinian approach to genetic selection. "There are skulls from 2000 B.C. that show the pulling apart of the skull from the expansion of the cortex." Today, 1 in 8 Sardinians has the mutant gene. If two gene carriers have a child together, there is a 25 percent chance the child will be afflicted. The child will have life-threatening anemia, necessitating transfusions and therapy to remove excess iron from the body.

Dr. Cao has been screening people since 1978 to determine if they are carriers of the thalassemia gene, and offering them prenatal diagnosis and abortion if their fetus is affected with the disease. Because his program has operated for two decades, it provides a glimpse into the effect of testing on people's lives that could help guide the United States in the inevitable design of programs to screen for the many diseases whose genes are currently being uncovered.

Dr. Cao attributes the success of his program to the fact that so many people in Sardinia actually know someone with thalassemia. "Since 1 in 8 people in Sardinia is a carrier," explains Dr. Cao, "in the small villages of 2,000 to 4,000 inhabitants, there are 2 or 3 patients."

This creates a high level of awareness, where women can make more informed choices about whether to undergo screening and whether to abort a fetus with thalassemia since they know what the life of such a child will be like. But it also can create more psychological anguish, since the woman may be symbolically aborting someone she knows.

In the United States, there are some similar instances, such as in families with cystic fibrosis. But a movement is afoot to test everyone for everything—to offer cystic fibrosis screening and breast cancer screening to the population at large, not just those with a family history of the disease. Given that a fetus can be tested for over three hundred disorders, women are forced to make decisions about whether or not to abort a fetus who has a condition with which they are unfamiliar. They can thus be unduly influenced by how the physician describes the disorder to them. But physicians may not be the best source of information about genetics.

A 1995 survey found that only 68 percent of the 128 medical schools in the United States required students to take a genetics course. In some of those schools, the "course" was only four hours long. Many new genes are being discovered each month and moved rapidly into use in clinical diagnosis without physicians being adequately prepared for advising patients about the appropriate circumstances for their use and the interpretation of results. In an assessment of the genetics knowledge of over 1,000 primary care physicians, for example, the average number of correct responses was only 74 percent. And according to another report, one-third of physicians erroneously interpreted the results of genetic testing for colorectal cancer.

Doctors' personal opinions influence the type of genetic testing they advocate; some cannot imagine life with a disability, so they urge patients to abort a fetus that may be impaired. When asked in a survey to evaluate their lives, 80 percent of the responding doctors said "pretty good," but 82 percent indicated that their quality of life would be pretty low if they had quadriplegia. In contrast, 80 percent of people with quadriplegia rated the quality of their lives as pretty good.

In Sardinia, there is a cute picture book, distributed to all students in the middle school, about a boy with thalassemia.

The walls of Dr. Cao's institute are decorated with posters and comic strips made by young students showing the inheritance mechanisms of the disease, the regions of the highest prevalence, and so forth. The posters sport slogans like NO ALLA TALASSEMIA CON UNA SCELTA INTELLIGENTE ("Say no to thalassemia with an intelligent choice").

Even in Sardinia, however, there is a problem with getting people to understand the manner of inheritance. Some individuals in Sardinia, who do not get tested, say, "I couldn't possibly have the gene; I'm blond." Still today, when a child is tested there and is found to have one copy of the mutant gene, each untested parent points to the other and says, "It must be from your side of the family." When a child has two copies of the mutant gene and is affected with thalassemia, however, the finger-pointing can't occur, because it means that each parent contributed a mutant gene. Psychologist Maria Luisa Palomba points out that "there are few divorces of couples with children with thalassemia. The men and women feel they have the same guilt in the same way. They can't divide it."

Of course, this is not true of all genetic diseases. Dominant ones, like Huntington's disease, are passed on by only one parent. X-linked ones, like Fragile X (causing mental retardation), are passed on only by the mother. There is already evidence from testing programs in the United States about the guilt and disruption this imbalance can cause. Women who pass on Fragile X to their sons feel an enormous guilt about failing in what they see as women's traditional role—having healthy males. And some people who learn they have a gene for a dominant disorder tell their spouse, "You can divorce me."

Palomba says that women who undergo prenatal diagnosis for thalassemia feel worse than women who undergo prenatal diagnosis for Down syndrome. "They have more problems because they feel that, while all women over a certain age are screened for Down syndrome, only those who are carriers have prenatal diagnosis for thalassemia. They feel guilty, because they are carriers. They feel they are marked."

Palomba found in Sardinia, as sociologist Barbara Katz Rothman found in the United States, evidence of a "tentative pregnancy." Couples often don't tell family or friends that they are pregnant until after the fetus has passed its test. When a Sardinian woman finds out the fetus is affected with thalassemia, if she already has an affected child, she asks permission of the first child to abort the fetus.

"It's as if they want to share their guilt with another person," says Palomba. "It's a symbolic abortion of the existing child. Usually, this only occurs if the existing child is fifteen or older, but there have been instances in which it has occurred with a child as young as the age of nine."

For siblings of affected people, there are different dynamics. "Women with affected brothers and sisters are more determined to have testing than mothers of such children or other pregnant women," says Palomba. "They know the problem. The healthy children suffer very much. They are marginalized from the family since all the attention goes to the sick child. They don't want to create a family where that will happen again. They feel less guilty, though, if their brother or sister is dead."

In Catholic Sardinia, voluntary abortion is not allowed for unwanted pregnancy with a healthy fetus. Palomba points out how the hospital staff may actually make the experience worse. She asks women to fill out a questionnaire, with one question asking their impressions of the abortion. Women often indicate that they were sabotaged by physicians or nurses, who treated them badly—refusing to change their IVs or help them if they are hemorrhaging—as a punishment for having chosen to abort an affected fetus. Since physicians are such respected authority figures, the women don't believe they can complain. "This thing usually happens in the night, when no family member is there," says Palomba. "You can't do anything against the doctors because no one can see you are telling the truth."

Although women in the United States are also sometimes made to feel guilty about their abortion decision (particularly if they must pass through a picket line amid placards of fetuses and endure hateful remarks to enter a clinic), there is much more pressure for women *to* abort in the United States than in Sardinia. Women are made to feel much more guilty here about the financial burden they are creating for society (or for other members of their health insurance plan) if they "choose" to have a child who can be predicted to require extensive medical care. Already, when an HMO paid for an amniocentesis of a woman that indicated her fetus had cystic fibrosis, the HMO informed the woman it would not pay for the child's care if she carried the pregnancy to term. In this way, genetic diseases become "pre-existing conditions" subject to insurance exclusion.

A woman I know was told by her obstetrician that her fetus had Down syndrome. The doctor ordered her to abort, she refused. She worked with disabled children and had no intention of terminating a pregnancy for what she considered to be a minor disorder. Another woman was similarly coerced. Her doctor told her that her baby would be more like a fish than a human and would only be as smart as a baboon.

A study by psychologist Theresa Marteau found that doctors blamed women who don't undergo prenatal diagnosis and then had a child with a genetic disorder. This is particularly troubling since physicians may be less likely to help women who decline testing because, says Marteau, "the outcome, giving birth to a child with a condition for which prenatal screening is available, is seen as preventable." Other research shows that physicians try not to spend time with patients when they believe the patients have contributed to the illness.

Margery Shaw, a doctor and lawyer, wrote in the *Journal of Legal Medicine* that women who carry a genetic defect such as Tay-Sachs and who go forward with a pregnancy of an affected fetus should be prosecuted as criminals—guilty of fetal abuse. She argues that "if parents know the genetic risk in advance, then failure to employ artificial insemination or in vitro fertilization could be considered a tort at the moment of conception." Shaw advocates a legislative scheme in which women with genetic defects would be required to use donated eggs if they wanted to have children. As gene therapy became available, that would be mandatory as well. Since all of us carry the genes for at least three to five lethal genetic defects, such a proposal could give the government the right to interfere in everyone's reproductive plans.

When I gave a speech at Emory Law School in Atlanta, law students came up to me afterward and said that, with genetic testing readily available, women shouldn't be allowed to have children with disabilities, since it raises people's insurance rates.

Nobel laureate Linus Pauling once suggested that people who were gene carriers of recessive genetic disease be tattooed on their forehead so they wouldn't have children with someone with the same gene. "If this were done, two young people carrying the same seriously defective gene in single dose would recognize this situation at first sight, and would refrain from falling in love with one another," wrote Pauling in the *UCLA*

Law Review. "It is my opinion that legislation along this line, compulsory testing for defective genes before marriage, and some form of public or semi-public display of this possession should be adopted."

Nancy Wexler, a psychologist at risk for Huntington's disease, bristles at the thought of these Scarlet Letters and asks how such a law would be enforced. "If economic sanctions are not invoked, what, realistically, are our options? Should gangs of medical thugs rove maternity wards looking for erring parents to sterilize?" she asks. "And how would the legislators determine what constitutes a sufficiently severe genetic impairment to be banned? Achondroplasia? Multiple polyposis? Schizophrenia? No balance of justice is so exquisitely sensitive as to calibrate life and death by statute."

Others envision that enforcement would come through mandatory prenatal screening with required abortion of affected fetuses. Margery Shaw even thinks the Catholic Church could be sold on the abortion of "defectives." She urges that Catholics just think of it as a permissible removal of an ill person from life support.

This is a personal matter for Marsha Saxton, a disability rights activist and lecturer at the University of California, Berkeley, who was born with spina bifida. "When I sit in meetings with public health officials showing slides indicating how much the state could save if pregnant women aborted fetuses with my condition," she says, "I have to keep saying to myself, 'They don't want to kill *me.*' "

Cao was quick to emphasize that, in his program, "we are not killing anyone. We are taking care of the whole problem." They do prenatal diagnosis and abortion, but in the same building they treat affected children with transfusions, bone marrow transplants, or a new experimental oral chelaton drug. They are also trying to learn more about how to change the genetic message that causes thalassemia in the first place. In fact, the Thalassemia Parents Association has an office in the institute. It raises money for treatment and gene therapy, but not for prenatal diagnosis and abortion—though some privately come for those services. In contrast, in the United States, once prenatal diagnosis and abortion became available for Tay-Sachs disease, physicians lost interest in finding a cure. A public health problem became "privatized" by asking women themselves to solve it through abortion.

Geneticist Angus Clarke points out that few of the health care professionals who offer prenatal diagnosis and abortion for particular conditions actually treat individuals with those conditions. "Whatever our personal feelings, our lack of involvement in such work must convey the impression that we think it less important than prenatal testing and secondary prevention," he says.

Genetic testing changes the very categories of "disabled." As bioethicist Paul Ramsey pointed out when amniocentesis was first introduced, "the concept of 'normality' sufficient to make life worth living is bound to be 'upgraded'" as testing is increasingly offered for less and less serious disorders. Currently, some parents choose to abort for reasons that seem trivial or inappropriate to other people. Some parents abort fetuses with an XYY chromosomal complement, for example, even though research has disproven the hypothesis that this is a "criminal" genome. People with dwarfism are enraged that someone would abort simply because the fetus would become a dwarf; in fact, some dwarfs seek prenatal screening to ensure their child is a dwarf, too. As time passes, an increasing number of women abort based on particular types of genetic information, which also changes the boundaries of normality. The percentage of women in Scotland who aborted fetuses with spina bifida rose from 21 percent in 1976 to 74 percent in 1985.

Newer genetic technologies provide the opportunity for users to choose to avoid very minor deviations from some perceived normality. Consider the use of preimplantation screening, which disability rights advocate Marsha Saxton refers to as "admission standards" for fetuses. A couple undergoes in vitro fertilization, produces multiple embryos, and each one is genetically tested. If a couple has ten embryos in petri dishes, they might use different criteria to determine the "genetic worth" of the embryos to be implanted than they would in determining whether a single in utero fetus at five months of development should be aborted. While a couple may not be likely to abort a fetus based on its sex or based on being an unaffected carrier of a recessive disorder, they may, when faced with a high volume of embryos, only a few of which can be safely implanted in the woman, decide to implant only noncarrier embryos or embryos of a particular sex. With ten embryos, the couple must refuse implantation of some of the embryos—for the safety of the woman and any resulting fetuses. If a couple chooses to move from randomly selecting the embryos to be implanted to doing so on genetic grounds, this may seem less morally problematic to them than aborting a particular fetus. The decision to choose the "best" embryos may be viewed differently than deciding about a developing fetus for several reasons. The woman is not physically pregnant, so there has not been attachment to a particular fetus. She has not felt the fetus move or begun to bond with it. In addition, the process may be viewed less as choosing *against* a particular individual (the developing fetus) than choosing *for* a set of individuals (the embryos to be implanted).

I was guided around the Sardinian institute by Dr. Maria Antonietta Melis. At the end of my visit, Dr. Melis invited me to a dinner at her home. She has a lovely apartment, with a large terrace growing olives, grapes, limes, tangerines, rosemary, and an assortment of herbs that she used in our salad and pasta. Once dinner was over, embarrassed, she asked me if she could smoke. She told me that she is made to feel like such a criminal in the United States about it that she doesn't smoke there. She said that she thinks it's ironic because the Americans were the ones who gave out so many cigarettes in Italy during the war; it's as if they taught the Italians to smoke. She also wondered why Americans were so harsh on people who do things that they themselves were doing a few years earlier, like eating meat or drinking alcohol.

I asked whether many children with thalassemia were put up for adoption in Sardinia (explaining that in the United States some children with AIDS or Down syndrome were), but she said no, that families here accepted them. Adoption occurred very rarely. She knew of only two instances in twenty years.

Dr. Melis described a woman physician who died a year or two earlier at age forty-five who herself was affected with thalassemia. I asked her if the medical school had suggested that she not apply because she would have a short life. I explained that this had happened in the United States with people who were at risk for Huntington's disease.

Melis was horrified. She said she couldn't imagine that happening, it was so cruel.

Then she asked why Americans were so—how do you say it, she asked, *intolerante?*

On the flight home, I thought about her question. Maybe the Italians can live better with the notion they are carriers of genetic diseases and can provide more resources to people who are born with disabilities because they have a better acceptance of fate than we do. Americans think they must be in control. They are used to it. Controlling their lives, controlling the world. A country that has directly suffered in war has less of that feeling of invincibility. Even the taxi system in Italy has the imprint of fate in the way an American one does not. When you call a cab there, you must pay whatever is on the meter when it arrives.

11

Gene Prospecting

❧

Eva (not her real name) has dark, close-cropped, elfin hair; big, alert, sparkling eyes; her neck is tilted to the left, and her arms and legs are rigidly bent at wrong angles. She looks like an off-centered marionette as she clops along in her laborious steps.

"They noticed it first in her eyes," Dr. Jack Penney tells me. "The old women of Barranquitas.... They said, *'Ella es perdida'* "—she is lost to the disease.

Eva can still walk and is greeted by much fanfare by the team of U.S. doctors and scientists who have come to Venezuela, as they do each year, to see her and the other members of her extended family. Jack whispers to me that she has the juvenile form of Huntington's disease, which is characterized by its rigidity; the adult form has more uncontrollable movements but limbs that, with effort, can be straightened momentarily by the doctor.

Jack starts the neurological exam on Eva. It is only my second day in Venezuela with a team investigating Huntington's disease, and I have the steps of the exam practically memorized. First, the patient follows your finger around with her eyes; with Huntington's, the command is not as readily received and so takes longer to carry out. Next, the patient is asked to point to her nose *(nariz)* and then hold out her finger *(dedo)*—back and forth as you say *"Nariz"* and *"Dedo."* There is a normal walking forward and backward and then doing so with one foot directly in front of the other. One part of the exam requires the patient to count backward from twenty with her eyes closed, and another calls for her to close her eyes hard for thirty seconds (*"¡Duro!, ¡Duro!, ¡Duro!"* Jack shouts—Hard! hard! hard!) and then *"saca la lengua"* (stick out her tongue). With superhuman effort, Eva is finally able to do it. ("That's what a strong will can do," Jack says later.) There's a part where the patient stands and the neurologist pushes her on the shoulders and asks her to try not to fall. Eva can't do this. She has to be caught after even the slightest push.

Over the next few days, I meet a dozen or more people with Huntington's disease. Some, like Eva, seem to be at the enchanted center of their families. Others are shut away. One patient has been locked alone in a one-room house with a large macaw in the yard. The patient's son lives next door, trying to convince himself his recent mood swings and somewhat jerky movements are not the onset of the disease.

None of the people I meet are patients in the traditional sense. None are in hospitals. That is not surprising since, in some of the places we go, our team is the only source of health care the people get. We give antibiotics to a grammar school girl with an infected leg wound, treat diarrhea, diagnose a slipped disk, and give vitamins to a pregnant teenager.

At one stop, I realize something is up when all three neurologists examine the same man. Later, Nancy Wexler tells me she could pick out the earliest signs of Huntington's disease when she first saw him through the window of his house. I can't tell. To me, he is a thin man in a yellow shirt who would be handsome if he did not have such a surprised look on his face. I take a photo of him before he is ready and get an even more surprised look—embarrassing. I sheepishly leave that one and take the better, second photo for the project.

"It's really sick," Jack says, "that I think it's a good day when I diagnose a new person with Huntington's."

• • •

Every March, Columbia University psychologist Nancy Wexler boards a flight for Venezuela. The striking, energetic blonde bypasses the shopping, beaches, and Latin sunsets of Caracas and Isla de Margarita, to enter a mosquito-laden village of tin huts and dirt floors. She visits a large extended family of 9,000 people, most under age forty, all descendants of a woman, Maria Concepción, who lived there in the late 1800s.

Nancy shares a bond with these people: she, like them, is at risk for Huntington's disease. In 1968, the year after Nancy graduated from Radcliffe, her mother started showing signs of the disease and two years later, attempted suicide. Nancy's uncle and grandfather had died young, although Nancy hadn't realized at the time that Huntington's disease was the cause.

Huntington's disease is a dominant genetic disorder. If a parent has it, there is a 50 percent chance the child will inherit the disease. It strikes in middle age, progressively destroying the brain's neurons, which retards one's motor skills and slowly causes confusion, irritability, insanity, and death. About 30,000 Americans have the disease, with another 150,000 at 50 percent risk of developing it.

Her mother's illness led Nancy to study psychology, writing her dissertation on the emotional impact of being at risk for a genetic disease. Nancy's father founded the Hereditary Disease Foundation to underwrite scientific studies of Huntington's disease. The research generally involved neurological studies, but in the late 1970s Nancy became increasingly interested in the possibility of using emerging technologies to locate the gene responsible for the killer. She read about restriction enzymes and decided to apply this new tool to try to find the Huntington's disease gene. Restriction enzymes—chemical scissors—cut people's long strands of DNA at specified points. People with normal genes will have the same length pieces when their DNA is cut. If there is a mutation at a site where a restriction enzyme normally makes an incision, it will not cut, and that individual will have a longer piece of genetic material.

It sounded simple, but the logistics were overwhelming. If you imagine a person's genome being large enough to circle the globe, then one gene would be about one-twentieth of a mile in length and a mutation in that gene could be just one-twentieth of an inch in length.

"Needless to say, several knowledgeable scientists told us that we were crazy to look for the gene in this haphazard, hit-or-miss fashion," says Nancy. "They predicted it could take fifty years or longer to find our target. What we were proposing was equivalent to looking for a killer somewhere in the United States with a map virtually devoid of landmarks—no states, cities, towns, rivers, or mountains, and certainly no street addresses or zip codes—with absolutely no points of demarcation by which to locate the murderer."

In 1979, using restriction enzymes to search for a disease was new and, says Nancy, "thought to be whimsical, if not heretical." Only one such enzyme had been discovered, and though many markers were known for other species, researchers were just beginning to search for similar chemical scissors in humans.

When thirty-four-year-old Nancy explained to researchers at the World Federation of Neurology meeting what she wanted to do, they spoke to her in the slow, loud voice that one uses with an errant child. They told her she didn't know what she was talking about, called her a "young girl."

She was undeterred. In March 1981, she made her first major research expedition to Venezuela with funds from the National Institutes of Health. She went to the *pueblas de agua*, towns built on stilts in the waters of Lake Maracaibo, where Maria Concepción had lived a century earlier. She tried to get the woman's descendants to give blood and skin biopsies to determine if genes of people affected with Huntington's disease differed from those without the disease.

No volunteer stepped forward.

"It's also in my own family," she told them. "My mother died of this disease and my grandfather and my uncles. I'm part of this same research. I've given blood, and I've given skin. But our family is too small to study, and the families in the United States are not as good as yours for this research. We really need your participation."

"I don't believe you," one of the people in the village said. They couldn't imagine someone from the wealthy, powerful United States having the same disease they had.

The group's nurse grabbed Nancy and paraded her around, pointing out her biopsy scar.

"If I had not had that biopsy scar, they might never have believed me," Nancy said. "It really made a difference. In Latin America, family was important, and they felt like I was part of their family. After that, even if they thought the whole thing was crazy, at least they never doubted our sincerity."

Nancy sent the samples she collected to Massachusetts General Hospital to be analyzed by James Gusella, a young researcher. He was taking a chance getting involved in such a study, because it might take years of frustrating dead ends—and no publications—before the gene was found. His career hung in the balance.

In 1983, Gusella hit the jackpot with just the third probe he used. It cut DNA so that the samples for people with Huntington's disease were a different length than those of their healthy relatives. The research team knew that the cut was near the Huntington's disease gene, and they could place it on chromosome 4.

This meant the team was getting close to finding the actual gene. Out of 3 billion possibilities, they now knew that the gene was on a particular stretch of 4 million base chemical letters on chromosome 4. "We had been incredibly lucky," says Nancy. "It was as though, without the map of the United States, we had looked for the killer by chance in Red Lodge, Montana, and found the neighborhood where he was living."

It took another decade, though, to locate the killer itself—to find the mutation in the gene that actually caused Huntington's disease. Nancy continued to go down to Venezuela for six weeks each March and April, and an ebbing cast of volunteers would join her for a week or two each. There was a master list of who was coming when. If supplies were needed, Nancy would call the next person in the United States who was due down.

The genetic mutation turned out to be a stutter in the genome. At some points in a person's genes a certain sequence of DNA repeats. Instead of having a particular three chemicals in a row once, they may be repeated a dozen times. With Huntington's disease the three chemicals (cytosine, adenine, and guanine) repeated anywhere from 40 to 125 times. The sequence of genetic CAG, CAG, CAG, in dizzying repetition, meant that those who had Huntington's disease—or would later in their life develop it—had a larger strand of DNA when cut by the chemical scissors.

Even when her team later successfully located the Huntington's disease gene—using the very technique the scientists had pooh-poohed—Nancy didn't get the recognition she deserved. "Oh, and by a little band of *amateurs*," one noted geneticist told her.

After the gene was found in 1993, the trips to Venezuela did not stop but rather took on more urgency. Nancy's group had never been content just to find the gene; they were after something much more important, a cure. For Nancy, approaching her fiftieth birthday, when Huntington's usually hit, the matter was intensely personal.

In March 1995, as I prepared to accompany Nancy on one of her trips to Venezuela, I received a memo warning me of some of the risks I would encounter: "Maracaibo and the barrio are quite close to the Colombian border. Although Venezuela does not have anywhere near as bad a drug problem as Colombia, what drug trade there is is often brought through Venezuela from Colombia across its borders, including Maracaibo. Reporters from the *Daily Mail* have been robbed and raped. The barrio in which we work is considered one of the most violent in all of Maracaibo. So BE CAREFUL!"

On one trip, the team had nearly drowned when their boat capsized in Lake Maracaibo. On another, there was a government coup and the team was restricted to their hotel grounds. (The hotel informed them of this by a note under their doors that said, "There is a curfew, but we still have Italian Fiesta Night.")

The group was an odd collection of workers, mainly doctors

from the United States donating their time. Some had been coming down for over a decade, ever since Nancy started the project. They had bonded with many of the at-risk people they studied, only to be devastated when the subjects ultimately developed the disease.

Nancy sat us down as a group before we hit the road. She explained about the project, about how one of the goals was to see if there was a correlation between repeat lengths and age of onset of the disease. She was also exploring whether repeat lengths expanded and contracted between generations, making it seem that the gene had disappeared from a family over several generations, when in fact it had just contracted, then re-expanded. We were trying to collect sperm (as well as blood) on this trip, a delicate prospect in a Catholic barrio. If individual men had sperm of different repeat lengths, maybe Huntington's disease could be "prevented" by choosing the particular sperm to inseminate an egg.

Nancy was also investigating the potential connections between obesity and the Huntington's disease gene. It was possible that overweight people get milder versions of the disease. There was no real scientific role for me with the group, but my Spanish was passable. My job would be to ask politely in Spanish whether I could weigh and measure each at-risk Venezuelan.

On my return trip, I would have one more duty. I would carry through Customs a liquid nitrogen tank full of the blood and sperm of at-risk men. It needed to be transported that afternoon to Gusella, so that it wouldn't arrive on the weekend when nobody would know how to deal with it.

I had a letter from the U.S. Department of Health and Human Services and the Venezuelan health minister asking

Customs agents in Miami not to open the tank, which would destroy the samples. When the time came, though, the agents weren't buying. I had flown into Miami on a flight that originated in Colombia, and now I was telling them they couldn't open the box I was carrying.

"Yeah, right," the agent said. He reached over to open the box, and I blocked his way.

I made him read the letter again.

"They're just Xeroxes," he said. "Your name's not even on them."

I started to explain the Huntington's disease project to him.

"Well, who are you?"

"I'm a . . . researcher," I said.

He gave one more suspicious glance at the box and then let me pass.

Before that trip home, I had a week of hot, sticky, heartbreaking days. At 6 A.M. on my second day in Venezuela, we made the two-and-a-half-hour trip to Barranquitas in our two vans. There were open ditches coming from the houses, filled with feces. We parked, grabbed the bags of supplies, and headed down a dirt path to the house of a man bedridden with Huntington's who had been having diarrhea for the past few days. The "house" was two adjoining tin rooms. It was 97 degrees outside, probably 110 degrees in the tin house as we all crowded in. There were about fifteen kids with us in the eight-foot-by-four-foot room, as well as the wife of the man who lived in the smaller adjoining room.

Nancy called me into the sick man's room. There was a clothesline string across, with everyone's clothing on it, covered with a little plastic—although it seemed to do little against the dust. "I want you to see how these people live," she said.

I counted the hooks on the wall for hammocks. Seventeen people slept in hammocks there—three or four to a hammock.

Hours and hours passed as the doctors took blood and did neurological exams to see if any of the people at risk had developed the disease since their last visit. After I weighed and measured the appropriate people, I alternated between the stifling heat inside the house and the blazing sun outside.

The children outside tried earnestly to talk to me, but I didn't always understand. They were fascinated by the large vaccination mark on my sleeveless arm. I saw a girl with polio, her one leg bent at improbable angles, and mentally went through my schedule for the two weeks after my return, trying to figure out when I could get my son Christopher his vaccination booster shots.

One girl pointed out which children had *el mal*—Huntington's disease—in their families. The cute little boy with the buzz cut. The girl who looked like a Gauguin painting, with her long brown hair, big eyes, brown shirtless skin, and bright purple skirt. She continued pointing. There were more and more. At 50 percent risk.

Everywhere we went, we saw malnourished children with bloated bellies and thin legs. They didn't go to school. They would begin having their own children, one after another, when they were thirteen. It made them a perfect group to study genetically—so many subjects for the research, like mice in a lab. But it also made the twenty-year-olds appear to be fifty. Even if a treatment was developed through this research for Huntington's disease, I wondered how they would be able to afford it. They didn't even have access to drugs for the much more common diarrhea that claims so many lives in countries such as this.

We drove around a few neighborhoods and parked on a

street. It was nearly noon. Marshall, Shelly, and I were in the van, and Nancy was questioning passersby, searching for other individuals who might be descendants of Maria Concepción. We decided it was a good time for lunch in the van. We had these fluffy, sweet, cornmeal pancakes, which we wrapped around white cheese. A police car pulled up across the street, one house up. A cop with a shotgun got out. His backup arrived, jumping out, finger on his shotgun trigger, and the two men began to charge the house. Nancy continued obliviously to talk to the citizenry of this barrio. Shelly and I noticed that our companion van, which had been parked in front of us, directly across from the SWAT team, was ever-so-gingerly backing up past us, parking half a block away.

Nancy was unfazed. She had found a man (who might some-how be at risk for Huntington's), and she talked him into the backseat with Shelly and me to get his blood drawn. She told him everything would be fine; he was with two *rubias* (blondes), Shelly and me. He was a little drunk and had been betting on the races, which blasted out over the radio of the nearby cantina. It formed a noisy incongruous backdrop to the police raid, as if I was watching one channel on the television and getting the sound from another. At one point, they came into sync. I mis-took the shouts and screams from the race for noises from the raid.

We spent the day tracking down leads, making U-turns, and following any vague suggestion from a passerby that there might be a distant relative with Huntington's disease only a few more miles down the road if we just followed some labyrinth-like instructions that he or she would give us.

At one of our stops, a team member took me aside. He was troubled by an encounter he had had two days earlier, when someone whose blood was being drawn had said, "You always come down here and take our blood, but you never tell us the results. How do I find out the results?"

He had then called Nancy over. "What results do you want to know?"

"I want to know if I will get Huntington's disease."

Nancy had explained how the study called for the investigators and patients to be blinded to any individual's genetic status, so it wouldn't influence their assessments of the patients.

The team member told me how patronizing he thought this was. The patient hadn't chosen to be blinded. In the United States, there would at least be a duty to refer to some places for the testing if the person wanted it. The research group treated the Venezuelans with compassion and respect. Yet there were tensions like these throughout the endeavor. Did the eight-year-old who signed a consent form really know what he was agreeing to? Would the research be conducted differently if it were being done in the United States?

Nancy's work in Venezuela translated into a test that people could take in the United States, Canada, and elsewhere to find out whether they had the mutant gene. The test was not offered in Venezuela. "We felt that if we gave diagnostic information and then left for a year, we would be acting like hit-and-run drivers," said Nancy. She pointed out that since abortion is illegal in Venezuela, the information can't be used in prenatal diagnosis.

One night we visited a grade school that was honoring Nancy. The students had collected money and food to give to the people in one of the barrios with a high incidence of Huntington's disease. The children in the class—a few of whom were themselves at risk for getting the disease—had written essays about how they would feel if they learned through genetic testing that they would get Huntington's disease.

"Then I wouldn't be able to be a teacher," one wrote.

In that one sentence, all the negatives of genetic testing came into focus for me. People who learned they would die young from a genetic disease could give up on themselves, be stigmatized, not live out their full potential. Imagine an eight-year-old giving up on her career goal because she might die four decades later. The disease could be cured by then.

The people of Venezuela, like the canaries in the mines, were the harbingers of what was to come with each new gene discovery. Some could not get jobs when they moved to cities because it was now known that people from Lake Maracaibo were at risk for the mysterious disease. People who were healthy said they would not marry people with the Huntington's gene.

The impact of genetic testing on people in the United States, Canada, and England was troubling as well. "Suicide is now a question of when, not if," one woman wrote to a psychologist after learning that she had the genetic mutation associated with Huntington's disease. The suicide rate is four times higher among people with Huntington's disease than among the general population. Even when people are prepared for bad news as a result of testing, they may still be shocked by the reality of it. For example, a woman who said before undergoing predictive testing for Huntington's disease that she expected to have the faulty gene nonetheless stated after she received the results, "I feel like someone has died. Part of me has died, the hopeful part."

Originally, Huntington's disease researchers expected everyone who tested negative for the genetic mutation—who learned they would not fall ill—to be elated. But Nancy soon discovered that even when the results of genetic testing reveal that a person does *not* have a genetic mutation, the results may cause psychological harm to him or her. Nancy called it "survivor's guilt," as they wonder why they have been spared when other family members tragically have inherited the gene.

Of people who undergo genetic testing for Huntington's disease and learn that they do not have the gene, 10 percent experience severe psychological problems as a result. Many people with a parent with Huntington's disease assume that they have inherited the gene. They may live their lives as if they will die of the disease in their fifties. Learning they do not have the gene also radically changes their self-image. One woman said, "If I'm not at risk—who am I?"

One man, who was at 50 percent risk for Huntington's disease since one of his parents had it, lived his life in preparation for his early death. He enjoyed dangerous hobbies, like skydiving. He spent money, rather than saving it, and ran up huge loans and credit card bills. He did not commit to a long-term relationship with a girlfriend because he did not want to get married and risk passing on the gene to a child.

Then he was tested and learned he did not have the Huntington's disease gene. He was just an ordinary Joe, who might need to start thinking about a mortgage rather than living hard, dying young, and leaving a beautiful corpse. The test results precipitated a downward spiral, in which he embezzled from his company to pay his bills.

And Huntington's disease was just the beginning. Then came tests for breast cancer, colon cancer, Alzheimer's disease, homosexuality, risk-taking behavior—all of which brought similar unsettling results.

What will be the impact on a married man who believes himself to be heterosexual when he learns he has a gene that is linked to male sexual preference? What about those individuals who discover they have a gene supposedly linked to aggression?

Genetic testing of a fetus can have a large impact on the parents' emotional well-being and self-concept. If the fetus is being tested for a dominant disorder, for which one of the parents is at risk, the diagnosis of the fetus with the genetic mutation means the parent has it as well. This comes up most strikingly with Huntington's disease, which does not generally begin to affect people until after age forty.

Most people who have a parent with Huntington's disease, and thus are at 50 percent risk of having the disorder themselves, decide not to get tested. Only 15 percent of at-risk people decide to take the test. They see no benefit in learning their status since no treatment is available. However, some who refuse to test themselves nonetheless test their fetus. Nancy observes that "if the test of the fetus for Huntington's disease is positive, two massive losses are suffered simultaneously." The couple will generally abort the fetus. And the parent will learn for the first time that they have the genetic mutation and thus, says Nancy, "hear their own death knell with that of their child."

Preimplantation genetic testing on IVF embryos offered new choices to such at-risk individuals. Doctors could test that person's embryos, discard affected ones, and only implant those without the gene. That way, the at-risk parent would not need to find out if he or she carried the Huntington's gene.

An Institute of Medicine committee I served on met to assess the ethics of the procedure. It seemed fair, compassionate, and practical.

But what if the doctor learned that all the embryos were okay, and that the woman did not carry the gene? Would the doctor be ethically obliged to inform the patient, so that she wouldn't bother to go the $16,000-plus expense and physical and emotional risk of IVF the next time she got pregnant?

"I would want to know if that was the case," said Ed Wallach, the former president of the American Society of Reproductive Medicine. He seemed oblivious to the fact that if 50 percent of the couples were told to create their second child the old-fashioned way, other parents using preimplantation screening would realize that they carried the gene. The elaborate preimplantation procedure would not have protected them.

People who learn through genetic testing that they have genetic mutations related to late-onset disorders may lose their job or insurance as a result, like the stigma of being from Lake Maracaibo. Even their untested family members may be discriminated against. Among people in families with a known genetic condition, 31 percent have been denied health insurance coverage of some service or treatment due to bad genes in the family. When Colorado schools started testing children for the genetic mutation called Fragile X syndrome (which causes mental retardation), some parents lost health insurance not only for affected children but for themselves and their healthy children as a result. And some people who participated in genetics research lost their insurance as a result, including a man who enrolled in a study of the colon cancer gene.

Vindictiveness against people with genetic mutations does not stop there, though. Other social institutions want to know a person's genetic profile as well. A medical school denied entrance to a man who was at risk of Huntington's disease. What's the value of training a doctor, they reasoned, if he would die young? And, in 1997, researchers suggested that boxers with the APOE 4 gene mutation should not be allowed to fight, based on loose evidence that head injuries in gene carriers increase the risk of Alzheimer's disease.

Courts are beginning to make decisions based on people's genes as well. At the behest of an ex-husband, a South Carolina court ordered Huntington's disease testing on the ex-wife, hinting at a future day when custody battles would be determined based on genetic tests predicting how long each parent would live rather than an assessment of the quality of relationship with the child. Rather than risk learning that she had a devastating, untreatable disorder, the woman in South Carolina disappeared to avoid the test—even though this meant she would never see her children.

As such examples mounted, Nancy Wexler went to Congress, urging that testing for Huntington's disease and other genetic maladies be voluntary and that protections be instituted against genetic discrimination. She showed that deciding whether to use genetic services involves more than a bit of tissue, a sliver of DNA. It involves the whole person, with impacts that touch all aspects of that individual's life.

Genetic information should remain private, she insisted. There are risks, as well as benefits, to gazing into that genetic crystal ball. Bioethicist Arthur Caplan of the University of Pennsylvania agrees—and suggests a radical solution—stopping all genetic testing until protective laws are on the books.

When *60 Minutes* did a story on the quest for the Huntington's disease gene, Diane Sawyer asked Nancy if, when she started her research, she had expected that she would take the test once she had discovered the gene.

"Absolutely," replied Nancy. "Yes. I never doubted it. And now I'm not sure."

12

Genetic Politics

∾

W hen Congress funded the Human Genome Project in 1990, James Watson, its first director, knew that we were embarking upon a giant social experiment. The $3-billion project was established to determine the location of and to analyze the constituent parts of each of the 50,000 to 100,000 genes in each human cell. The ultimate goal is to facilitate the development of genetic diagnostic tests and genetic treatment modalities for the nearly 4,000 diseases that have a genetic basis. But Watson took an unprecedented step: he set aside 3 to 5 percent of the budget to fund studies of the ethical, legal, and social implications of genetics.

The ELSI effort, as it was called, was established to serve as a watchdog for the Human Genome Project, to hold it accountable, and to protect the public. It granted money to researchers outside the National Institutes of Health to study problems associated with genetics. (I received a grant, for example, to study the psychological, social, and legal implications of genetic testing.) And it sponsored the ELSI Working Group to advise the National Institutes of Health, Department of Energy, Congress, the president, and the public at large about how to prevent harms from genetic technologies.

Many in the scientific community were outraged that Watson was using part of the scientific budget to police them. Watson told them they couldn't avoid it, that "the cat was out of the bag" with respect to the public's concerns about the impact of genetics.

"Yes," one NIH researcher replied, "but why *inflate* the cat? Why put the cat *on TV*?"

In three years, Watson built the new Human Genome Project at the National Institutes of Health, not by establishing a new lab there, but by awarding millions of dollars in funding to genetics researchers all over the country. Those scientists—and researchers at other institutes within NIH—were raising ethical issues through their work. Dean Hamer at the National Cancer Institute had discovered a "gay" gene, which he claimed predicted sexual orientation. Should parents test fetuses for the gene? And while the Armed Services could fire gays, the Department of Defense had adopted a "don't ask, don't tell" policy. Now the Armed Services were realizing they didn't have to ask; they could just "peek"—by analyzing the blood samples they had on hand from every member of the military. Could a soldier be discharged if he had the gay gene?

J. Craig Venter, a researcher at NIH's National Institute of Neurological Disorders and Strokes, raised the level of controversy to a new order of magnitude. At the start of the Human Genome Project, determining the sequence of a whole gene, often made up of tens of thousands or more genetic letters, was still a cumbersome project. One of the breast cancer genes has 100,000 pairs of letters. Frustrated with the slow pace of identifying all the DNA letters in a gene, Venter came up with the idea of spotting new genes by looking at just part of the sequence. Instead of manually sequencing DNA, he started putting it through a high-speed, computerized sequencing machine.

At the time, no one was quite sure how partial gene sequences might be used. But in the genetics gold rush of the early 1990s, Venter's boss, NIH director Bernadine Healy, wanted to make sure that Venter and the NIH held the property rights to these sequences. Previous generations of biological scientists had been inspired to do research in order to gain scientific knowledge, as well as rewards of scientific prizes and tenure. But under the Republican Congress, new laws were passed, saying that publicly funded researchers could now patent their findings and could enter into joint ventures with biotechnology companies. The potential windfall for scientists was huge. Consumer groups bridled at the thought, annoyed that they would have to pay twice. Taxpayer dollars funded the NIH research, yet if a gene was patented from the research, any company using the gene in a diagnostic test or a gene therapy could be required to pay a patent royalty, increasing the cost of the product to the consumer.

That indeed happened with AZT, the anti-HIV drug. Developed and tested with public funds from the National Cancer Institute, the drug's patent rights were given to a pharmaceutical company, Burroughs Wellcome. This was in sharp contrast to the policy of the March of Dimes, which prohibited patenting or the receipt of royalties for the research projects that developed the polio vaccine. When Jonas Salk was asked who would control the vaccine, he said, "Well, the people, I would say. There is no patent. Could you patent the sun?"

I wondered how human genes could even *be* patented. Patent law covered "inventions" and prohibited patenting of "products of nature"—a tree or a rock or an eagle. It also prohibited patenting scientific formulas, like Einstein's $E = mc^2$. Genes seemed to be both a product of nature and a formula.

In July 1991, at a congressional briefing on the Human Genome Project, Venter dropped the bomb that NIH had filed patent applications on the gene sequences he had isolated from human cells. The patent traffic was likely to be huge; Venter's technique enabled him to find small identifying sequences of 2,000 genes per month. Senator Al Gore opposed NIH's patent grab, pointing out it was causing other governments and private researchers to patent as well, since NIH's patent filing was "universally viewed as an attempt to corner the market on human genetic information."

James Watson denounced the idea of patenting partial gene sequences as "sheer lunacy," saying that with the new automated sequencing machines, "virtually any monkey" could do what Venter's group was doing. A prime tenet of patent law was that the "invention" had to be nonobvious and useful. Partial gene sequences were neither. Watson insisted that the important job was in the interpretation of the sequences, not the sequencing itself.

What most irritated Watson and others was the notion that by simply sequencing a short piece of an unidentified gene with an automated sequencing machine, a researcher could lay claim to the entire gene. This would undercut patent protection for those who are performing the real work of elucidating the function of the proteins encoded by the genes, thus discouraging further research.

Watson's opposition to Healy's plan to patent the gene

sequences became increasingly public. They quarreled, and on April 10, 1992, Watson quit the Genome Project.

The search for a person to replace Watson as the head of NIH's National Center for Human Genome Research quickly zeroed in on a University of Michigan researcher, Francis Collins. His impressive credentials included being part of the team that discovered the Huntington's disease gene. In 1989, he had been a codiscoverer of the cystic fibrosis gene—and had immediately filed for a patent on it.

In April 1993, Collins moved to the National Institutes of Health. As part of his deal, he negotiated the creation of his own laboratory at NIH, something Watson had not done. Collins was in charge of a program that would give millions of dollars to university genetics researchers. But he would oversee their proposals and perhaps be able to use information in them to his own competitive advantage.

Collins, a devout Christian, was an immediate hit in Congress, attending prayer breakfasts with some of its conservative leaders. He even offered a divine warrant for what he was doing.

He cited the Bible, Matthew 9:35, as justification for this massive project to map the human genome. "Jesus went through all the towns and villages, teaching in their synagogues, preaching the good news of the kingdom, and healing every disease and sickness," he wrote in 1997. "It is part of our mandate as Christians to pursue such medical advances, attempting to emulate Christ in his healing role."

Collins sided with the Church when it was useful to his gene-hunting efforts, but scolded it when religious leaders voiced a broad objection to the patenting of genes on the principle that it was an attempt to patent God's creation—life.

"As a Christian, it was a black day for me when the church

that I care so much about supported a statement [opposing the patenting of life forms, including genes] that was considered ludicrous in the scientific community," he wrote. Collins pressed for religion to be more rational about the issue of patenting. "To the extent that we speak with reason and love, God is glorified. If we do not, he is not."

But the commercializing of genetics was changing the fundamental nature of science. A 1996 study, by Tufts University professor Sheldon Krimsky, of 789 biomedical articles published by academic scientists from universities in Massachusetts, found that in 34 percent of the articles at least one of the authors stood to make money from the results they were reporting. They either held a patent or were an officer or adviser of a biotech firm exploiting the research. *None* of the articles disclosed the financial interests of the authors.

Surveys find that patenting has led to reductions in openness and data sharing, delays in publication, and tendencies to select research projects of short-term commercial interest. In several cases, corporations with vested interests had tried to suppress the publication of research findings that were not in their interests. Strains over conflicting commitments have caused some researchers to sever their commercial ties.

J. Craig Venter, the gene sequencer, left NIH in 1992 to join a commercial venture called Human Genome Sciences (HGS), which established a nonprofit research branch, the Institute for Genomic Research (TIGR) for Venter to head. The following year, HGS signed a $125-million pact with the pharmaceutical giant SmithKline Beecham giving it commercial rights to the genes Venter discovered.

Geneticist David King described the situation this way: "You have a corporation trying to monopolize control of a large part of the whole human genome, literally the human heritage."

Venter eventually soured on the commercial constraints on his scientific research and split from HGS. He was not amused when he published an article estimating, based on his latest scientific research, that there were 60,000 separate human genes, and a venture capitalist from HGS had berated him: "What the hell do you think you are doing saying there are only 60,000 human genes? I just sold 100,000 to SmithKline."

Pressure arose elsewhere. British doctors doing cystic fibrosis testing on children balked at having to pay royalties to the American owners of the cystic fibrosis gene—Francis Collins or whoever he assigns his patent rights to. The British newspaper the *Independent* described the situation thusly: "You may think that your body is your own. But at least one in 20 of the people reading this article possesses a gene that a team of North American scientists is claiming as its own."

As Leon Rosenberg, while dean of the Yale University School of Medicine, put it: The influence of the biotechnology revolution on medicine has "moved us, literally or figuratively, from the classroom to the boardroom and from the *New England Journal* to the *Wall Street Journal*."

I have seen this trend firsthand. When I attended a Human Genome Project conference at Stanford, the auditorium was packed with scientists when George Poste, the scientific director of SmithKline Beecham, spoke. Poste, whose company was making multimillionaires of many genome scientists, talked about the benefits of genome research.

The crowd of scientists thinned out when Nancy Wexler stepped up to the podium to address the ethical and social issues raised by the Human Genome Project. I myself had to go out in the hall to return a phone message marked urgent. As I made my call, I realized that the very scientists who needed to hear Nancy's talk had left the auditorium and were relaxing on couches in the hall. As I dialed, I could overhear their conversation. One of the older ones, maybe thirty-five years old, was advising a younger one how to remain calm if the biotech company shares he had been given for his work began to drop.

• • •

When John Moore, a patient with hairy cell leukemia, had his spleen removed at the UCLA School of Medicine, he discovered just what this attention to commercialization meant in terms of patient care. Moore says his physician patented certain chemicals in Moore's blood, without his consent, and set up contracts with a Boston company, negotiating shares worth $3 million. Sandoz, the Swiss pharmaceutical company, paid a reported $15 million for the right to develop Moore's cell line.

Moore began to suspect that his tissue was being used for purposes beyond his personal care when UCLA cancer specialists continued to take samples of blood, bone marrow, skin, and sperm for seven years. When Moore discovered in 1984 that he had become patent number 4,438,032, he sued the doctors for malpractice and property theft. His physicians claimed that Moore had waived his interest in his body parts when he signed a general consent form giving the UCLA pathology department the right to dispose of his removed tissue. But Moore felt that his integrity was violated, his body exploited, and his tissue turned into a product: "My doctors are claiming that my humanity, my genetic essence, is their invention and their property. They view me as a mine from which to extract biological material. I was harvested."

The trial judge in the Moore case threw out the lawsuit on the grounds that the body parts could not be considered property, but an appellate court was more sympathetic. This higher court reviewed cases involving celebrities such as Bela Lugosi, who was held to have a property interest in his likeness that prevented other people from marketing photos of him. "If the courts have found a sufficient proprietary interest in one's persona, how could one not have a right in one's own genetic material, something far more profoundly the essence of one's humanity than a name or a face?" What's more, since organ donation laws give patients control over what is done with their bodies after they die, it seems logical they should have control *before* they die.

Moore's doctor and UCLA appealed the case. On behalf of the People's Medical Society, a patients' rights group that did not want to see patients' bodies turned into a treasure trove for their doctors, Berkeley law professor Margery Shultz and I filed an amicus brief in the case, arguing that Moore's doctor should have told him what he planned to do. We argued that people have a right to know if their doctors have a conflict of interest in using the patients' bodies for commercial gain.

The California Supreme Court agreed with our argument and said that doctors must inform patients in advance of their surgery when their tissue would be used for research or commercial products. But it denied Moore's claim that his cell line was his property. The court decided that the doctor and biotechnology company rather than the patient should profit from Moore's unique cells. The decision rested on the promise of biotechnology innovation. The court was concerned that giving Moore a property right to his tissue would destroy the economic incentive for biotech companies.

Justice Allen Broussard wrote a scathing dissent. "Far from elevating these biological materials above the marketplace, the majority's holding simply bars *plaintiff*, the source of the cells, from obtaining the benefit of the cells' value, but permits *defendant*, who allegedly obtained the cells from plaintiff by improper means, to retain and exploit the full economic value of their ill-gotten gains free of their ordinary common law liability for conversion."

UCLA, a defendant in the case, had raised visions of a brave and troubling new world. It had argued that even if Moore's tissue was his property, since UCLA was a government institution it would be entitled to take the tissue against his will under "eminent domain."

The *Moore* case was decided in 1990. By 1998, though, even venture capitalists were beginning to question the grant of patents for human genes since licensing fees must be paid to the gene "owner"—the holder of the patent—by later researchers who develop truly useful items such as a diagnostic test or gene therapy related to that gene.

In May 1998, the journal *Science* published an article showing how patents can deter innovation in biomedical research. "A proliferation of intellectual property rights upstream may be stifling life-saving innovations further downstream in the course of research and product development," wrote Michigan law professors Michael A. Heller and Rebecca Eisenberg.

They likened the situation in genetics to that of postsocialist economics. The expectation in eastern Europe was that private stores would be loaded with goods once a free market was introduced. But the stores remained bare—while street vendors flourished. The reason: no individual could set up shop without collecting rights from workers' collectives,

privatization agencies, and local, regional, or federal governments. Similarly, with genetics, wrote Heller and Eisenberg,
"privatization can go astray when too many owners hold rights
in previous discoveries that constitute obstacles to future
research." Since there are more than one hundred patents,
for example, related to the adrenaline receptor, a researcher
whose work is related to that site faces a daunting bargaining
procedure.

• • •

The ELSI Working Group had been created to cover troubling
social issues like the commercialization of genes and insurance
discrimination based on genetics. Nancy Wexler was the first
chair. When her five-year term ended in 1995, I was elected to
replace her.

The transition from a psychologist as a leader to a lawyer
seemed logical. Social science research had showed us the problems that genetics raised; now it was up to law to do something
about it. But could a watchdog group that was housed within—
and funded by—the NIH's National Center for Human
Genome Research actually turn a critical eye to what the scientists there were doing?

In many scientific quarters, the Working Group's concern
with the psychological and social impact of genetic testing was
seen as too touchy-feely. The acronyms given to the efforts
didn't help much. The mostly male international group of
genetic researchers were known by the male name HUGO
(Human Genome Organization). The Working Group chaired
by first one woman, then another, was called ELSI (the Ethical,
Legal, and Social Implications of the Human Genome Project).

"Before you take the position, make sure they fund a staff person at your office," Nancy advised me, "rather than at the Genome Center at NIH."

But I assumed that the Genome Center staff and I were in this together. I saw no problem with using National Center for Human Genome Research employees in Bethesda to provide the day-to-day staff work for the committee. After all, their boss, Francis Collins, cared passionately about genetic discrimination. He told me that when he was searching for a cancer gene, he lay awake some nights wondering if people would be better off knowing they would fall ill later in life. He worried that insurers would use that information against them.

I was so pleased to be working with Francis Collins to create genetics policies that I overlooked the fact that he had begun to stack the ELSI Working Group by adding as a voting member a genome scientist, whom the Working Group had not nominated nor voted on, thus violating the ELSI charter. I decided not to challenge the fact that he pressured me into having my first Working Group meeting be on a topic of his choice, even though we were supposed to be able to set our own agenda.

As months passed, I noticed a disturbing trend. The ELSI Working Group was given no budget of its own; instead, we had to ask Collins and his staff for funding. Every time we planned an activity that might lead to more people getting genetic tests or participating in genetic research—such as a project protecting genetic privacy—we were given a blank check. But each time we planned an activity that called into question the power of genetic testing—such as a study of the problems with using genetics to predict intelligence, criminality, or certain psychiatric disorders—we were told that the Genome Center didn't have enough money to fund it.

When *The Bell Curve* was published, a book that suggested that African Americans were genetically intellectually inferior to whites, a philanthropy journal warned that donors might be wasting their money trying to change black children's lives through educational programs since inherited intelligence "appears to have an enormous effect on what a person is able to do, from the ability to earn income to the likelihood of becoming a criminal or a teenage mother."

The members of the Working Group felt it was inappropriate for the Human Genome Project to stand by and let the 400,000 readers of *The Bell Curve* believe that blacks were genetically inferior to whites. So we used scientific data to show that people are not just a pattern of genes unfolding; the environment makes a difference. "Since the lessons of genetics are not deterministic, they do not provide useful information on deciding whether or not to pursue various programs to enhance the capabilities of different members of society," we concluded. "Those decisions are moral, social and political ones."

We had asked our staff director at the Genome Center to mail out the statement to scientific journals. Months passed, and she failed to do so. We asked her repeatedly, but she always had an excuse. She was "too busy" (even though she was supposed to be working exclusively for us). She also said she thought publications wouldn't be interested.

The statement was getting stale. Months had passed since the book's publication. Finally, we forced our staff person to send it out, and it was immediately published in *Science* and written about in *Nature*.

The Human Genome Project was creating numerous societal dilemmas—issues of commercialization and the disproportionate impact of genetic testing on women and minorities—but these sorts of questions were not getting proper attention. So much of NIH's ELSI money for outside researchers was going into studies of how to test for two of the genes Francis Collins had worked on—the cystic fibrosis gene and the breast cancer gene—that at one ELSI meeting, a chart was distributed showing only *one* study had been funded that grant cycle that was not on one of those two topics.

Rather than put the "cat on TV," as the NIH researchers had initially feared, my new position was keeping this cat out of the media. Shortly after I became chair, the *New York Times* called me for a comment on the problems with breast cancer testing. I never returned the call. I didn't want to offend Collins, a breast cancer researcher, and I didn't want to make it appear that I was too opinionated to run the committee.

In September 1995, Francis Collins held a massive press conference at NIH to announce that he and colleagues from NIH, Hebrew University, and the University of California at San Diego had found a mutation in the breast cancer gene, the 185delAG mutation, that had disproportionately occurred in Ashkenazi Jewish people. His study, however, had found only *eight* people with the mutation. By holding a big press conference on the subject, he was putting news of the discovery on the front page of the *Los Angeles Times* and the *Washington Post*. Within days, women were calling their doctors asking to be tested. One clinic even began offering to test prenatally, so Jewish women could abort fetuses that might get breast cancer later in life.

The press conference had been held before the necessary epidemiological studies were done to see what the chances were that women with the mutation would actually get breast cancer. The doctors who began testing women told them there was an

86 percent chance they would develop breast cancer. Some women had both their breasts removed in an attempt to avoid that high risk. When the appropriate studies were later done, it was found that only 50 percent of women developed cancer. Some women had amputated their breasts unnecessarily.

Collins had looked for the mutation using blood samples that Ashkenazi Jewish couples had submitted to their doctors for clinical screening for the Tay-Sachs or cystic fibrosis gene, so that they could make their own reproductive decisions. They hadn't given their blood for breast cancer testing.

There was nothing illegal about doing research on those samples after the reproductive carrier screening was over, since Collins and his colleagues removed patient identifiers. But it seemed to some people, including myself, unethical not to ask for people's consent to do research on their blood. Not all the Ashkenazi Jewish women may have wanted to participate in such a study, particularly since its results might be used by insurers to charge their group higher rates for health insurance or might reawaken old eugenics ideas that somehow Jews were "genetically inferior."

Also in September 1995, ethical concerns were raised about the other agency we advised, the Department of Energy. That month, African American employees at a Department of Energy lab—the Lawrence Berkeley Laboratory at the University of California—sued the lab for undertaking genetic testing on them without their consent. The ELSI Working Group had been advising DOE for five years on the ethics of genetic testing, for which the first tenet was voluntariness. And now DOE employees were saying the agency tested them without their knowledge or consent.

The suit claimed that when African American employees

underwent annual physicals at the Lawrence Berkeley Laboratory, their blood was surreptitiously tested for sickle-cell anemia carrier status. Women at the lab were given surreptitious pregnancy tests, and all employees undergoing physicals were secretly tested for syphilis. The test results, placed in the employment files, were not disclosed to the employees.

A forty-six-year-old African American administrative assistant at the lab, Vertis Ellis, found out that at each of her six company physicals over her twenty-nine-year period of employment, she was tested without her knowledge or consent for sickle-cell anemia, syphilis, and pregnancy. Mark Covington, another lab administrative assistant, says he tested positive for sickle-cell trait but was not told of the results.

Yet when I questioned DOE officials about the suit, they said that since there was no proof that they discriminated against people based on genetic test results, they did nothing wrong by roaming around in employees' genes.

What good was an ELSI Working Group if it didn't even have an impact on the primary agencies it looked after?

As we started asking tough questions of our federal sponsors, they made it increasingly clear they did not want to hear from us. At our October 1995 meeting—a month after Collins's press conference and the DOE lawsuit—the ELSI Working Group raised concerns about the marginalization of ELSI. The National Center for Human Genome Research told us the budget was so tight it would have to cancel two of the next three meetings and renege on the $20,000 it had promised us for a project investigating genetic tests by courts, schools, and other social institutions. The Working Group vice chair, Berkeley sociologist Troy Duster, was irate. In addition to sitting on the Working Group, he was a member of the National Advisory

Council for Human Genome Research, the NIH committee that funded genome scientists. He said he couldn't reconcile NIH's cutting the $20,000 to ELSI activities with the "$12 million to this lab, $8 million to this lab" discussions that he routinely sat through on the council, where grants were given to geneticist friends of council members.

When it came to funding molecular biological scientists, says Duster, the council was willing to take risks and give millions to laboratories that were using speculative, high risk, and unproven techniques that had not in the past been as successful as had been expected. At the same time, several vocal and influential members of the council were often unsupportive of and trivialized the research on social, legal, and ethical issues, begrudging the 5 percent spent. But what he found most troubling, Duster says, was his inability to reconcile the millions poured into speculative techniques at the molecular level, while the Council claimed it could not find funding for even a full series of ELSI Working Group meetings that year.

When I later got the minutes of the October 1995 meeting, prepared by Genome Center staff, they bore little resemblance to what had happened. The minutes were drafted to make the Working Group look bad and the Genome Center look good. I asked the staff person who drafted the minutes to send me audiotapes recording the meeting, so that I could redraft the minutes. She said she would mail them. Weeks passed, and they didn't arrive. I called again. She said she would give them to me when I came to NIH in December. When I asked for the recordings in December, she told me she would mail them. The tapes still haven't arrived.

I met with Francis Collins to talk about topics for the next meeting. I suggested the Working Group hear from some of the psychologists the ELSI program had funded to assess the psychological impact on people of learning they had the gene for a disease that would affect them later in life.

"Too mushy," he said.

Then I said I wanted to put gene patents on the agenda.

"No," he said, "HUGO is looking into it." But HUGO, that group of international genetics researchers who would benefit from getting patents, did not seem like the best venue for the full range of social concerns to be honestly assessed.

The Institute of Medicine of the National Academy of Sciences had said that the ELSI Working Group must be independent so that it does not "give the impression of the fox guarding the chicken coop." The NIH itself had promised Congress that the Working Group would have autonomy. As NIH director, Bernadine Healy had told Congress that ELSI "is an independent group that is examining legal and social implications of human genome research, but it is outside of any particular ideology. It is quite removed and quite independent."

In December 1995, I went to Washington to complain to Francis Collins that our "autonomy" was becoming laughable. He stood me up to attend a prayer breakfast with a member of Congress.

In February 1996, I quit.

I went quietly, refusing to talk to any reporters about the issue. I stood by while Collins painted the dispute to the media in whatever terms he liked. In part, I was afraid to take on NIH, with its enormous power and $13.6-billion annual budget. From my work as a lawyer, I knew how rough it was to be a whistle-blower.

I was touched, though, by the number of supportive e-mails I received, including some from genome scientists. "Oversight groups are supposed to be a pain in the ass and are probably not doing their job appropriately if they are not," wrote one scientist. "Francis should know that. Not everyone has to agree with you and you shouldn't get your way all of the time, but the group is necessary to assure that appropriate concerns are raised, important questions are asked and discussed, and that there be a publicly recognized route to monitor and question the development and application of a pervasive technology."

Even though I had been raising my concerns for months with him—as had other members of the ELSI Working Group—Collins called me and claimed he was "shocked" by my decision. Then he asked if I would be willing to stay as chair until May, so that it wouldn't interfere with his upcoming Human Genome Project budget hearings in Congress.

He still didn't get it.

But then, in order to deflect attention from my criticisms, he decided to impanel a committee to investigate my resignation. While his staff had told the ELSI Working Group that there was only enough money to meet one more time in the next ten months, suddenly money was found for the other group to meet numerous times—not to solve any of the major policy issues genetics raised—but to assess the ELSI Working Group's relationship with NIH.

In December 1996, when it was likely to get the least media attention, Collins released the investigatory group's report. The report substantiated my concerns, pointing out that the Working Group's autonomy had been compromised by being under the thumb of the Genome Center: "Serious concerns have been raised about the Working Group's lack of resources and independence" and "the inadequate sharing of information by the staff."

The investigatory group recommended formation of a new Working Group, not staffed out of the genome center, but instead staffed out of Secretary of Health and Human Services Donna Shalala's office to remove it from the conflict of interest.

The rest of the existing Working Group quit, to make room for the new and improved Working Group. But Secretary Shalala did not appoint such a group, leaving Collins free and unencumbered in doing what he wanted.

His Genome Center was elevated to a full-fledged institute at NIH, with a bigger budget and more power.

Then, in May 1998, Craig Venter entered into a joint venture with Perkin-Elmer, a company that makes superfast gene-sequencing machines. Venter would, using 230 of Perkin-Elmer's machines at once, finish sequencing the human genome in three years—cheaper, sooner, and better than the federally funded Human Genome Project could do it. His announcement caused Congress to ask why it should continue to fund the federal project if a private company was going to beat the government labs to the punch, particularly since none of the six sequencing centers the government has funded under the Human Genome Project had so far achieved the sequencing rates the six centers promised two years earlier. This has led to these government-funded sequencers being jokingly referred to in the scientific press as the Liar's Club.

Francis Collins had the most to lose from Venter's announcement, since his budget might be cut substantially if Venter's project went forward. He scrambled to lobby Congress and the public on the necessity of continued funding to his institute (including a ten-page *New Yorker* ad, which included a picture of Collins on his motorcycle, and opened with a line about how excited a young girl was when Collins and his group found a "jean" for cystic fibrosis). In June 17, 1998, testimony, Collins told Congress that NIH could do a better job at sequencing than Venter and would "discourage patenting."

"A concern that has been raised in many publications is how the intellectual property issues associated with generating the entire human genome will be handled," Craig Venter told Congress, when it was his turn to speak. "By making the sequence of the entire human genome available it makes it virtually impossible for any single organization to own its entire intellectual property."

Rather than Collins's weak (and not-so-credible) statement that he would "discourage" patenting, Venter promised bluntly, "Our actions will make the human genome unpatentable."

Collins did not reveal that he himself patents genes, which can result in an extra $150,000 per year to him under congressional and NIH rules. Two weeks after the hearing, he and two colleagues received a patent on the gene for ataxia-telangiectasia, a gene discovered with NIH funds. Since it can predict certain cancers, it is a potentially lucrative discovery.

No one is paying attention to the fact that the free-enterprise policies that Collins is developing as head of the Genome Project will benefit him when he leaves NIH. His Department of Energy counterpart, David Galas, has already cashed in, by leaving the government agency to become president of a biotechnology company, Chiroscience R and D, Inc.

Even if Venter succeeds in making the entire gene sequence per se unpatentable, other ethical, legal, and social issues related to genetics will remain. There will be questions of the commercialization of genetic tests, the appropriateness of genetic engineering of humans, the use of genetic information by social institutions, the problems of behavioral genetics. Yet there is no longer an independent ELSI Working Group to address these issues.

The need for an outside monitor of NIH is great. "When the National Institutes of Health makes an announcement through one of its many spokespeople, who checks out the credibility of that statement?" asks Kary Mullis, the 1993 Nobel laureate in chemistry who invented PCR (polymerase chain reaction), the technique to amplify DNA that is essential to many genetic tests. "Checks and balances are hard to come by in a scientific establishment that is supported by a populace unskilled in scientific arts."

When it comes to NIH, even the Fourth Estate doesn't do its usual investigations. Lisa Belkin, who covered for the *New York Times Magazine* Venter's new quest to scoop Francis Collins, did not mention Collins's patents in her article, even though one of the criticisms he makes is that Venter will commercialize the genome. And, when she quoted a supposedly neutral scientist, Eric Lander, she failed to point out that he received over $10 million from the Federal Human Genome Project.

Nor were the members of Congress that useful when I appeared at their request to discuss genetics issues. I had dealt with Congress in the mid-1980s on reproductive technologies and gene therapy, but when I led a briefing for congressional staff on genetic discrimination a decade later, I was surprised at how things had changed.

I told them about real people who had lost their jobs or their insurance because genetic tests predicted they would develop a late-onset illness.

An aide to one of the congressmen responded, "We're not interested unless you get us a celebrity to talk about it."

At a genetics policy meeting, I learned why James Watson

had formed the ELSI Working Group. Watson implied that the ELSI Working Group had been created not to set ethical standards but to let the science proceed unimpeded.

"I wanted a group that would talk and talk and never get anything done," Watson said, "and if they did do something, I wanted them to get it wrong.

"I wanted as its head Shirley Temple Black."

13

More Oversight for a Tattoo

∽

There are two brothers, identical twins, both air traffic controllers. One brother wants Huntington's disease testing; the other does not. The first brother says that he'll keep the results secret from the second. But will he actually be able to do so?

If the first brother doesn't have the gene, won't he want to tell his twin the joyous news? And if he doesn't relay good news to his sibling, won't his brother then know that they both have inherited the gene? If the first brother's medical record indicates that he has the Huntington's disease gene, the second, untested twin could be denied health and life insurance based on those results—and both may lose their jobs as air traffic controllers.

I tell this story during grand rounds at a teaching hospital, and after my talk, some of the doctors and lawyers from the hospital take me to lunch. The geneticist is beeped and goes off to return the call.

When he returns to the table, he says, "I have an ethical dilemma for you. A man just called and said he was holding a

pair of his girlfriend's undergarments in his hand and wanted
to have them genetically tested to see if the semen on them
was his."

"Did he say he was holding a large-caliber automatic weapon
in the other hand?" I ask. Then, after a pause, I say, "What's
going to happen if the sperm isn't his?"

"That's none of my business," the doctor says. "I just provide
a service."

He explains that men often bring in their children and, for
$1,000, obtain tests to determine if the children are biologically
theirs. "How is this any different?" he asks.

I make a mental note to find out if the rate of wife abuse has
gone up in the neighborhood surrounding the hospital. But
those sorts of possibilities don't concern him. Rather, he is worried that the woman might have some legal claim against him
because the panties are her "property."

The internist sitting next to me has a worried look on her
face. She turns to the hospital attorney. "Am I on the same malpractice insurance carrier as he is?"

• • •

Everywhere I look, new reproductive and genetic technologies
are being offered, without sufficient thought about their impact
or desirability. "We can tell you how many swabs are used on
animals in a year in this country, but we can't tell you how many
people were involved in fertility research in this country or how
many adverse events there were," says Wisconsin law professor
R. Alta Charo. "We require all of that for nonhuman animals,
but not for people."

"A woman gets more regulatory oversight when she gets a

tattoo than when she gets IVF," says Brooks A. Keel, professor of obstetrics and gynecology and associate dean for research at the University of Kansas School of Medicine in Wichita.

My phone rings daily with calls from journalists, judges, and government agencies, wanting my opinion on the latest technology. GIFT, ZIFT, ICSI, the number of acronyms grows. The Japanese have tried this, the South Koreans that. I feel as if the world is locked into a battle over who can push the boundaries the farthest.

As I look over my phone messages, I think of the prank calls teenagers often make. Some days my messages look equally whimsical. Human sperm in mice. Male pregnancy. One temporary receptionist at my office threw out half my phone messages. She thought they were a joke.

In the course of my career, I have learned several truisms: If it has worked in just one animal, it will be tried in a woman. If a baby is born from the technique, her picture will go up on the clinic wall, but no one will study how she fares as she develops, nor how her mother does over time.

Sometimes I am taken aback thinking of the 300,000 IVF children that have been created. None of them has yet tried to have a child of his or her own. What if they have trouble reproducing, as did the children of women who had been given diethylstilbestrol (DES) during pregnancy? Will I have been complicit in their misery?

And what about the women? They are trapped in an endless cycle of trying. Just when they think they are ready to come to terms with their infertility and remain childless or adopt, their doctors offer a new acronym. Those baby photos on the clinic wall stare down at them. "Quitter . . . failure," they seem to chant. So the women give it another go.

When I first entered the field, in vitro fertilization was used only on women with hopelessly damaged fallopian tubes who needed it to circumvent their infertility. The doctors told me it would be *unethical* to subject a fertile women to the risk of hormonal stimulation, retrieval of eggs, and so forth, to address her husband's infertility. Instead, the couple could create a child through artificial insemination by donor, a widely accepted, safe procedure at a cost less than a tenth as much as IVF, with a much higher success rate and less risk to women.

But that moral boundary did not last for long. In 1993, doctors began offering ICSI—intracytoplasmic sperm injection—to couples where the husband had a low sperm count. Even though the wife didn't need it, she was put through all the rigors of IVF so that her eggs could be harvested and directly injected in vitro with her husband's sperm. Where, previously, a man was considered infertile unless he produced millions of sperm per ejaculate, now a man can be fertile even if he can produce only a single sperm.

For the doctors in the lab, the procedure itself was thrilling. They took a thin needle and shot the sperm into the egg. It was like sex under the microscope.

There was in fact something so satisfying about controlling conception that the doctors started using ICSI even in cases where the husband produced enough sperm to fertilize an in vitro egg without the injection. Within four years, more than one-third of all IVF procedures involved ICSI. Tens of thousands of children were born after ICSI.

In Belgium and Australia—unlike the United States—the government keeps track of how many children conceived through reproductive techniques have genetic abnormalities. In 1998, researchers in those countries noticed that the children created by ICSI were twice as likely to have major chromosomal abnormalities as were children conceived naturally. At age one, the children of ICSI showed developmental delays in problem-solving ability, memory, and language skills compared to children conceived through sex or IVF.

When ICSI is used because the father has inheritable infertility, his sons will be infertile and will need to use ICSI to reproduce as well. So, with each procedure, the clinic is creating a new generation of clients.

ICSI is not the only technique being used without sufficient advance study. The lack of regulation of fertility and genetic techniques means doctors can introduce experimental procedures into their clinical practice soon after they are invented, often long before they have been adequately researched in animals.

While frozen sperm and frozen embryos have routinely produced pregnancies, it has been difficult to figure out how to freeze eggs. In 1986, an Australian doctor produced the first known births from frozen eggs; a year later, German doctors produced frozen-egg pregnancies. However, physicians in the United States were unsuccessful at producing frozen-egg births or pregnancies.

In 1994, Reproductive Biology Associates (RBA), in the Atlanta suburb of Dunwoody, began running an experimental donor frozen-egg program. After several years of trial and error refining their technique, RBA doctors asked a thirty-nine-year-old IVF patient if they could inseminate a frozen donor egg with her husband's sperm. The patient, who reportedly could barely afford her treatments, was told that if she agreed to enroll in the experimental donor frozen-egg program, she would not be charged for the treatment.

RBA doctors thawed twenty-three eggs from a twenty-nine-year-old donor and used the woman's husband's sperm to fertilize the sixteen eggs that survived the thaw. They used ICSI to fertilize these eggs. Eleven embryos resulted, and four were implanted into the woman—resulting in twin boys in August 1997, making her the first American woman to give birth to a baby born from a frozen egg.

But in November 1997, South Korean researchers published a study in *Fertility and Sterility* that suggested eggs frozen at an early stage of development and then thawed showed an increased incidence of chromosomal abnormalities compared to fresh eggs. The researchers suggested that the freezing itself damages the DNA inside the eggs. Joe B. Massey, RBA's director, admitted that this study made him rethink using frozen eggs to create embryos. Then in May 1998, RBA said that it hopes to routinely offer egg freezing within the next five years, and the Jones Institute for Reproductive Medicine, at Eastern Virginia Medical School, announced in September 1998 that it had begun to offer egg freezing to its patients.

In 1998, doctors at Joseph Schulman's Genetics and IVF Institute clinic removed pieces of the ovaries of a twenty-year-old cancer patient, Stacie McBain. The clinic charged $11,000 for the procedure. Doctors froze the pieces for later use in creating eggs, after chemotherapy left McBain infertile. "They said they'd used it in sheep and it worked," McBain told *Washington Post* reporter Rick Weiss.

Weiss notes that McBain was not told, however, that only *one* sheep had ever become pregnant through this method. A brochure that the clinic provides refers to the "extraordinary results" in sheep. Yet patients are required to sign a consent form that says in part "we cannot guarantee that you will benefit" from the procedure.

For men who don't produce sperm, Ralph Brinster of the University of Pennsylvania Veterinary School suggests taking immature cells from the man's testes and maturing them in a pig's or bull's testes, which would then produce human sperm. Other researchers suggest a more promising alternative to the artificial womb: gestating human fetuses in animals of a compatible species, such as cows.

"Part of the way we think about who we are and how we value ourselves has to do with our origins and reproduction," says bioethicist Arthur Caplan. "Something is challenging the specialness of humanity if you originate human beings in some animal's reproductive tract."

When the problem is with the woman's egg rather than the husband's sperm, an egg donor can be used. But donated eggs from women volunteers are scarce, with prices escalating up to $5,000 per egg. So a search has begun for alternative sources of genetic material to produce babies.

Researchers are focusing on the biological fact that all of a woman's eggs are created prenatally. A woman actually has her maximum number of eggs—about 7 million—when she is a twenty-week-old fetus in her mother's womb. By puberty, only around 300,000 remain, and up to 400 will be ovulated before she reaches menopause.

With a million abortions a year, some scientists have begun to think the unthinkable—using female fetuses as a source of eggs for infertile women.

Immature eggs can be harvested from an aborted fetus and matured in a laboratory. The eggs can then be used in a regular in vitro fertilization process. Another variation is to transplant pieces of a fetal ovary into a woman whose ovaries are not working or are not present. A British scientist, Dr. Roger Gosden, set off an international furor at the beginning of 1994 when he announced that he had accomplished this procedure in mice and that it could be done in humans within three years.

A woman who receives a fetal ovarian transplant may be at risk of contracting diseases, as can happen with the transplants of any tissue and blood. Fetal tissue has to be frozen and tested at three-month intervals for HIV. If this step is skipped, the recipient risks contracting AIDS. And, when eggs are taken from fetuses and not subjected to natural selection forces (which cause some eggs to ovulate, while other eggs expire and are reabsorbed into the ovary), there may be some risk of abnormality in the children. As seen with ICSI, bypassing natural selection can have disturbing results.

While some infertility doctors saw nothing wrong with fetal eggs, the public felt otherwise. In the British Parliament, Baroness Strange said, "I don't believe that, like rag dolls, babies should be made up of discarded bits of humanity." She was joined in this sentiment by Baroness Ryder of Warsaw, who accused medical researchers of turning reproduction into a "manufacturing process."

Concerns were raised about the psychological impact on the child, who some said would grow up knowing that "his grandmother murdered his mother." How would the child feel? While learning that her genetic mother was a dead fetus might not be the same type of loss as the death of a mother she knew and loved, a child might nonetheless grieve. She might feel angry at her biological grandmother for consenting to the abortion or receiving money for the eggs, which prevented her mother from living.

Others counter that while the idea is troubling at first glance, the reasonable child will come to realize that life is better than no life. A *British Medical Journal* article reported that using fetal eggs is "no more yukky to the average ten-year-old than sexual intercourse." But we do not usually look to ten-year-old children for an assessment of the morality of an issue. And while most children presumably become comfortable with the idea of sex as a means of procreation as they mature, the fact that many adults still feel the "yuck factor" when thinking about the issue of fetal ovarian transplantation indicates that the concept is not a comfortable one.

And then there is the next logical extension: a woman who cannot produce eggs could be cloned, carry the cloned fetus for a few months, then abort it to remove its ovaries. Her IVF doctor would harvest eggs, which would be identical to those the woman's ovaries would have produced. The doctor would then fertilize the eggs from the abortus with her husband's sperm in vitro to create the children she and her husband would otherwise have had.

Yet there is a complex concern about what this line of research says about females. The process, says lawyer Adrienne Davis, transforms fetuses from a mass of cells to "gendered being." They become girls who would have grown up to become women and who would have been able to make their own reproductive decisions such as whether to procreate or not. The harvesting of eggs might move us closer to "fetal farming," possibly turning fetuses into a source for spare body parts.

In Britain, one member of Parliament called the practice "grave robbing," another "Orwellian practice." In 1994, Parliament made it a criminal offense to use fetal eggs or ovarian tissue for fertility treatment. Although no such law has been passed in the United States, infertility providers are not yet offering the procedure. Perhaps the yuck factor is influencing the practice. Or maybe it is just a matter of time. In some

circles, fetal eggs are being touted as safer for the resulting children than a woman using her own eggs. Anthony Smith, author
of *Sex, Genes, and All That, the New Facts of Life*, opines that fetal
eggs are better than women's since they have not been exposed
to certain environmental or chemical factors such as "natural
radiation, disease, fever, medication, [and] occasional excesses
of . . . alcohol." Some scientists see it as a logical extension of
their current advice that older women use younger donors' eggs.

Also pushing social boundaries is the possibility of male
pregnancy. A man could be primed with an injection of female
hormones, and then an in vitro embryo could be inserted into
his abdominal cavity. A placenta would develop and, with luck,
attach to the omentum, a fatty, blood-rich tissue that hangs in
front of the intestines. Nine months later, the baby could be
extracted in a procedure akin to a cesarean section.

Cecil Jacobson, the researcher who used his own sperm
to create babies for women who thought they were getting
sperm from anonymous donors, reports that when he was at
George Washington University, he transferred a baboon embryo
into a male baboon and let it gestate for five months. Even in
human women, pregnancy outside the uterus occurs in one in
10,000 pregnancies. Most do not succeed—and, indeed, can be
life-threatening to the woman and the fetus—because they
occur in the cramped quarters of a space like the fallopian tube.
But some women do deliver such babies with surgical intervention, including a woman who got pregnant after her womb was
removed in a hysterectomy but whose abdomen nourished a
five-pound baby girl.

Dr. David Kirby at Oxford University tried to achieve male
pregnancies in mice by implanting embryos in the testes. One
developed perfectly for twelve days, but that was only half the

normal gestation period. The testes were not big enough to accommodate the growing fetus.

The logic of male pregnancy seems quite similar to that of having the wife of an infertile male undergo in vitro fertilization because of her husband's infertility. In such a case, the woman undergoes hormonal stimulation and other procedures to avoid using a third party, the sperm donor. With male pregnancy, if the woman has an infertility problem preventing gestation, her husband can carry the fetus to avoid the use of a surrogate mother.

"If we are truly a society based on constitutional equality, regardless of race, religion or sex, then that is exactly what should happen," one doctor wrote to me, emphasizing that the right to use reproductive technologies should belong to men as well as women.

Others disagree. "Today, at least, the attainment of pregnancy is not something that any sane man would attempt, or that any ethical physician would suggest," wrote Princeton biologist Lee Silver in his book *Remaking Eden: Cloning and Beyond in a Brave New World*. This is an odd position for an advocate of all sorts of cutting-edge reprogenetic technologies—ICSI, genetic engineering of embryos, even cloning—that present risks to women and children.

The fact that most researchers and doctors are male obviously influences the technologies that are available. Men don't even have to take fertility drugs: it's the women (and the children) who are at risk with most reproductive technologies.

Boston University health law professor George Annas suggests that a new agency be created to review what he calls "boundary-crossing experimentation." This would include human cloning, genetic engineering, organ transplants from animals to humans, and artificial hearts. Just as the FAA regulates aviation, this new agency would regulate reproductive and genetic technologies to protect the consumers.

In the United Kingdom, a national licensing authority was established under the Human Fertilisation and Embryology Authority. No new technique can be tried without HFEA's approval. When such an oversight group has been suggested in the United States, reproductive technologists have argued that they should not be singled out for regulations that do not apply to other areas of medicine. Yet the constraints usually in place in other fields of medicine are lacking here.

Unlike new drugs and new medical equipment, which are regulated by the Food and Drug Administration, no similar review of innovative reproductive technology procedures is required. Reproductive technologies also differ from other medical procedures because they are rarely covered by health insurance; only a dozen states' laws mandate infertility coverage. For other types of health services, health insurers, through managed care outcome studies and evaluation of services, have required certain proof of efficacy before medical services are reimbursed.

Additionally, medical malpractice litigation, which serves as a quality control mechanism in other areas of health care, does not work as well in this field. The normal success rates for the procedures (25 percent for in vitro, for example) are so low that it makes it difficult to prove the doctor was negligent. Risks to the children may not be discernible for many years, which may be past the period of time a statute of limitations on a legal suit has run. In "wrongful life" cases, courts have been reluctant to impose liability upon medical providers and labs for children born with birth defects where the child would not have been born if the negligent act had been avoided; only three states recognize such a cause of action. Consequently, experimental techniques are rapidly introduced in the more than three hundred high-tech infertility clinics in the United States without sufficient prior animal experimentation, randomized clinical trials, or the rigorous data collection that would occur in other types of medical experimentation. This is truly the Wild West of medicine.

Should it be up to individual doctors to decide which new technologies should be used to create the next generation? In other areas of medicine, that is not the case. Most medical research in university and other hospitals is initially funded by the federal government through the National Institutes of Health and, by law, must be reviewed in advance by a neutral committee, an Institutional Review Board, before it can be tried in humans. Since reproductive technologies have been held hostage to the abortion debate, they have not received federal funds. Researchers can still submit their plans to hospital and university Institutional Review Boards, but they usually do not. In fact, according to IVF doctor Mark Sauer, IRB review of reproductive technology proposals is so rare as to be "remarkable." Even those rare studies that go before IRBs are not assessed for their social impact, however. The federal regulations covering IRBs specifically state that the reviewing committee should not address the social advisability of the project. The law says, "The IRB should not consider possible long-range effects of applying knowledge gained in the research (for example, the possible effects of the research on public policy) as among those research risks that fall within the purview of its responsibility." There is nothing akin to the override in the European patent law for procedures that are against morality or harmful to the public order.

Joseph Schulman did submit his experiment on ovarian

tissue freezing to the Institutional Review Board at Inova Fair-fax Hospital. But the chair of the IRB, Peter Paganussi, says that they knew Dr. Michael Opsahl at Schulman's institute would offer the procedure no matter how they decided. "The Genetics and IVF Institute already had run a series of ads in the *Post* announcing this service," the IRB chair wrote to the *Washington Post*. "Our feeling was that if we approved his study, at least we could monitor his actions and collect meaningful data about the safety and efficacy of the procedure."

• • •

Reproductive technology is tougher to regulate than nuclear technology. The tools for reproductive technology are relatively inexpensive and widely available. "A reprogenetics clinic could easily be run on the scale of a small business anywhere in the world," notes Princeton biologist Lee Silver. There are IVF clinics in at least thirty-eight countries, from Malaysia to Pakistan and Thailand to Egypt.

In the United States, the assisted reproductive technology industry, with an annual revenue of $2 billion, is growing to serve the estimated 1 of 6 American couples of reproductive age who are infertile. Annually, in the United States alone, approximately 60,000 births result from donor insemination, 15,000 from IVF, and at least 1,000 from surrogacy arrangements. In contrast, only about 30,000 healthy infants are available for adoption. What is so striking about this comparison is that every state has an elaborate regulatory mechanism in place for adoption while only three states, Florida, Virginia, and New Hampshire, have enacted legislation to comprehensively address assisted reproductive technologies. And they aren't even the states where the most high-tech reproduction is conducted!

"What is it going to take before we get some regulation in this area?" asks Arthur Caplan. "We've already had an untold number of women undergoing 'selective reductions,' a fifty-five-year-old woman on welfare who had multiples, and an unmarried man who contracted to have a surrogate bear his child and then killed the baby. If that hasn't prompted regulation, what will?"

"This country is the only country in our technological position that hasn't, as a society, faced up to the various social and ethical issues involved in this technology," says Harvard law professor Elizabeth Bartholet. One of the issues we haven't faced is whether we should, as did Great Britain, ban some reproductive technologies.

"The great challenge to mankind today is not only how to create, but to know when to stop creating," said Lord Emmanuel Jacobovitz, former chief rabbi of Britain, when he heard about the possibility of using fetal eggs. "And, when we celebrate a Sabbath to remind ourselves that G-d initially created this world, we celebrate not his act of creation on the six days. We celebrate that he knew when to stop."

14

The Sperminator

⌘

A few years ago, an elite Midwest teaching hospital called me with an unusual question: "We have six men in comas whose wives, girlfriends, or parents want their sperm. What do we do?"

Boarding the plane to the hospital, I thought about John Irving's novel *The World According to Garp*, in which the central character was the product of a liaison between a nurse and a comatose patient. The nurse in that case knew exactly what she wanted. But I was skeptical that the six women involved with this Midwest hospital had—independently—come to the conclusion that they wanted their loved one's sperm. I suspected that, lurking in the background, I would find an andrologist (a specialist in male infertility) who saw a medical journal article waiting to happen.

As anticipated, the physician in question already had his slides prepared for the talks he would give at medical meetings once he got the go-ahead to do the procedure.

This midwestern andrologist proposed to collect sperm from

comatose men in the same way it is collected from paraplegic men—through a technique called electroejaculation. An instrument that looked like a cattle prod would be inserted into the man's rectum. An electric shock would then cause an involuntary orgasm.

The andrologist then went on to tell me about his "patients." One was a twenty-five-year-old man who had been trying to father a child before a car accident put him in a coma with a poor prognosis for recovery. His wife wanted his sperm extracted.

Another was a forty-year-old man who also had sustained a severe head injury, also from a car crash. He had a child from a previous marriage, but none with his second wife, a woman in her midtwenties. He had told friends that he did not want to have additional children. The wife claimed that, in the week before the man's accident, he had changed his mind.

Although conception by men in comas was a new wrinkle in reproductive technology, I was already familiar with the idea of posthumous pregnancy. In 1983, I interviewed Kim Casali, the cartoonist who draws the syndicated "Love Is" cartoons. When her husband, Roberto, underwent chemotherapy for cancer, he froze sperm so that he could father a child if he recovered—or allow his wife to create a sibling for their existing infant if he died. After Roberto passed away, Kim successfully used the sperm. The birth announcement listed the parents as "Kim and Roberto (posthumously)."

In 1991, William Kane, a Yale-trained lawyer, froze fifteen vials of sperm specifically intending that his girlfriend, Deborah Hecht, be inseminated after he committed suicide. They chose a name for the baby, Wyatt, and William wrote a letter to the unborn child: "I have loved you in my dreams, even though I never got to see you born." Then he wrote a will giving the sperm to his girlfriend and saying about his impending suicide, "I'd rather end it like I lived it—on my time, when and where I will, and while my life is still an object of self-sculpture—a personal creation with which I am still proud. In truth, death for me is not the opposite of life; it is a form of life's punctuation." He took his life in October 1991 after bequeathing "all right, title and interest" in his sperm to Deborah, saying she should use it to become pregnant.

His two adult children by a previous marriage were not at all thrilled with the possibility of a baby brother or sister. Even though Deborah signed a release saying that neither she nor the resulting offspring would make an inheritance claim against the estate, the existing children—arguing through their attorney/mom, Kane's ex-wife—sued to have the sperm destroyed. The children, whose parents had divorced twenty-five years earlier, characterized their father's desire to create children after his death as "egotistic and irresponsible." They said destroying the sperm would "prevent the disruption of existing families by after-born children" and prevent "emotional, psychological and financial stress on those family members already in existence."

It seemed to me the argument proved too much. If accepted, it would mean that any firstborn, such as myself, could sue her parents if they depleted her resources (or future inheritance) by having another child.

The judge, though, accepted the kids' plea. Perhaps he was troubled by such an untraditional family. The mere fact that Kane requested this bizarre approach to baby-making seemed to indicate that he was not of sound mind. So the judge ordered the sperm destroyed.

Deborah Hecht appealed the decision—and won. The appellate court scolded the trial judge for his precipitous decision, saying that Kane "had an interest, in the nature of ownership to the extent that he had decision-making authority as to the sperm within the scope of policy set by law."

The appellate court asked the trial judge to reconsider the case. Since Deborah had inherited 20 percent of the estate, the trial judge gave her 20 percent of the sperm—3 vials.

Deborah was inseminated with William's sperm, to no avail. She sued again, and in 1997, when she turned forty-two, she won the rest of the sperm. As of late 1998, she was trying once more to get pregnant. Meanwhile, William's existing children have sued her again, this time for emotional distress.

Hecht was actually an easy case. William Kane clearly intended the posthumous use of his sperm in precisely the manner in which Deborah Hecht intended to use it.

A case in France was slightly more difficult. In that case, Corinne Parpalaix, joined by her in-laws, sued a sperm bank for access to sperm that her late husband, Alain, had frozen prior to cancer treatment. She said the sperm should pass to her as part of the estate, but the bank countered that "sperm is an indivisible part of the body, much like a limb or organ and is therefore not inheritable absent express instructions."

In the French decision, the court characterized sperm not as property or part of the person but as "the seed of life . . . tied to the fundamental liberty of a human being to conceive or not to conceive." The court focused its analysis on whether Alain "unequivocally" intended Corinne to bear his child. The court held that the evidence established his "deep desire" to make his wife "the mother of a common child."

Back at the Midwest hospital, the comatose men did not have wills stating they wanted to have children via electroejaculation. Nor was it likely that any had talked to their friends about that possibility.

In the first case the andrologist told me about—involving the twenty-five-year-old—there *was* evidence that the man and his wife had been trying to have children in the months preceding the accident. However, I had trouble imagining that a man's wishing to have children whom he would actually raise necessarily indicates consent to have children whom he will never see, touch, or interact with. In the second case, the man had told people that he didn't want children with his second wife at all.

The law is clear that if a woman wants an abortion, her husband can't force her to have a child. Why should she be able to force her husband to give her one? Some men in comas do wake up, after all. How would such a man feel to learn he had become a father through no effort of his own, while comatose?

The andrologist argued that doctors had been harvesting sperm from dead men for years. Dr. Cappy Rothman did so as early as 1978, retrieving sperm from an unmarried nineteen-year-old, for the benefit of his parents, who wanted to continue their lineage.

A 1997 survey found that fourteen clinics in eleven states had honored requests for sperm collection from the dead. Requests came from wives, girlfriends, and parents. The men, as old as sixty and as young as fifteen, had died of automobile or motorcycle accidents, lightning strikes, construction or farming accidents, or falls. The process was sufficiently commonplace that the American Society of Reproductive Medicine had a protocol, "Posthumous Reproduction," for dealing with it. Yet even though infertility doctors have been collecting and storing dead men's sperm for decades, few women were actually using it to procreate. It wasn't until May 1998 that a woman used sperm collected *after* her husband's death to get pregnant, in contrast to the more common use of sperm that the dead husband had frozen prior to his death.

California is the epicenter of dead men procreating, owing largely to Cappy Rothman. It's actually a sideline of his. As a urologist and andrologist, he has a one-third interest in the for-profit California Cryobank, which distributes 2,000 ampules of donor sperm a month to impregnate women whose husbands are infertile. In 1997, Rothman's donor sperm was used by women in forty-five different countries.

On Rothman's Web site, www.cryobank.com, you can search for a donor using categories such as religion, ethnic group, height, weight, whether there has been a reported pregnancy, eye color, hair color, hair type, blood type, occupation, and years of education. You can also pay $25 each to hear donors on audiotape.

Rothman also distributes T-shirts with attractive turquoise pictures of sperm on them and the words (in English or Japanese) *Future People*. It makes feminists furious, seemingly harkening back to Aristotle's idea that reproduction occurred when men deposited homonculi—the miniature essence of a human—into women, who contributed nothing but the flowerpot in which the seed could grow.

In October 1997, Rothman and I testified in a New York Legislature hearing where the question on the table was: Should a dead man's sperm be used without his previous consent? Rothman claimed there is less grief for the wife and other family members if some of the dead man's sperm is saved. He told legislators, "In one case where a man died by gunshot and I collected his sperm, his family followed me to the sperm bank and were consoled by seeing mobile sperm under the microscope. To console families in that way at a time of grief and tragedy is clearly part of my responsibility as a healer."

Yet different family members grieve in different ways. As the Kane case showed, some family members might not be pleased by the idea of fathering from the grave. And sperm wasn't the only source of dispute. A wealthy California couple, Mario and Elsa Rios, died in a plane crash with two embryos frozen in an IVF program in Australia and a multimillion-dollar estate. Would the embryos inherit the estate, or would the estate inherit the embryos? The answer to that question would make a world of difference. If the embryos were just property of the estate, the *existing* Rios children might choose never to implant them in order not to share the wealth.

On the off chance that the embryos might be beneficiaries, though, women began lining up in Australia to carry them. An Australian advisory committee recommended destruction of any embryos (including the Rios') where there was no advance directive about what should be done after the parents died. Right-to-life protests about "killing" the Rios' embryos, though, led to a reprieve. The Victoria, Australia, legislature enacted a law preserving the Rios' embryos for implantation. But since a California court ruled that the embryos would not inherit, far fewer women have been willing to act as the surrogate.

Cappy Rothman doesn't have a process for negotiating such disputes. What if the *wife* didn't want to use her dead husband's sperm, but the man's *parents* did, suggesting they would hire a surrogate mother?

"Deciding whose right to sperm should prevail is not my role," Dr. Rothman says.

But he also denies that role to bioethicists. "What makes an ethicist know what the rest of us should or shouldn't do?" he asks.

Existing law doesn't provide much guidance either. Under the Uniform Anatomical Gift Act—the organ donor law in each state—a wife or other relative can donate the deceased's organs or tissue, and can choose the recipient. Technically, then, the wife could donate the sperm to herself.

Or could she? The law covers transplantation, but using sperm to create a child doesn't quite fit that description. And wouldn't the wife have a conflict of interest? The law allows the wife to donate organs to further her husband's wishes, since it is unlikely that she would, say, need a kidney at the exact moment the man died. To remove the incentive for the wife to go against her husband's desires, payment for organs is not allowed.

But if she wanted a child and he didn't, the existing law wouldn't protect his desires after death. Claims of conflict of interest already are being leveled about fetal tissue transplantation. A federal advisory panel specifically recommended that a woman aborting a fetus *not* be able to designate a recipient even though she technically may do so as its next of kin under the organ donor law. The concern is that women will conceive and abort just to try to help out relatives, by providing fetal tissue, for instance, for treatment of Alzheimer's disease. Isn't donating sperm to herself just as bad? And why should a man's *parents* (the next in line under organ donor laws) have control of his reproductive capacity? There is no "right" to carry on one's lineage; if the son was competent and healthy, for example, his folks could not force him to have a child.

Even if the woman does not have the authority to direct the electroejaculation as a surrogate decision maker for the husband, can she make an even bolder claim and argue that his sperm is somehow *her* property (or at least their joint property)? A 1954 Illinois court case banned artificial insemination by donor on the grounds that it was adultery. The underlying theory was that each spouse has a claim to the other's reproductive capacity.

Although more recent court cases are uncomfortable with categorizing a man's sperm as his wife's property, another case, *Brotherton v. Cleveland,* held that a wife had a property interest in her husband's corneas. In that case, the wife had refused to donate organs or tissue from her dead husband because of his aversion to such a possibility. Nevertheless, the coroner retrieved Brotherton's corneas for donation. The wife successfully sued for damages. In that case, though, giving the wife a property interest was a way to effectuate the husband's prior wishes. In the coma cases, there is no evidence about what the men would do in these circumstances.

And unlike the French case where the sperm existed on its own, the coma cases would require andrologists to invade men's bodies to get it. (There is some appeal on equity grounds to seeing men as sperm containers, given that women in recent years have been treated increasingly like fetal containers in court decisions in which women are forced to have cesarean sections or other procedures for the benefit of the fetus.) However, with the sperm still in the men, rather than in the bank, courts are likely to find that electroejaculation violates their liberty interest in bodily integrity. The U.S. Supreme Court believes that a person in a coma or a persistent vegetative state continues to have such a right.

Once comatose men are turned into objects from which tissue can be harvested by their wives, what's to keep men from arguing for equal rights? If the wife were comatose, could her husband ask for eggs to be retrieved? Could he argue he had a constitutional right to impregnate her and keep her alive on a respirator for nine months until the child could be delivered by cesarean section?

In one case, a comatose twenty-eight-year-old woman was raped by an attendant at her nursing home. When nurses noticed that the woman's stomach was swollen, they gave her three pregnancy tests before telling her parents the truth. Her Roman Catholic parents refused to authorize an abortion. A son was born prematurely, weighing less than three pounds, and the nursing home paid the parents over $6 million to avert a lawsuit. The young woman died shortly before the boy's first birthday, but the child has survived.

Although couples can't force their children to give them a grandchild, new reproductive technologies allow them to take matters into their own hands. A twenty-three-year-old man who was planning to attend medical school was involved in a motor vehicle accident that left him in a vegetative state. The patient's father asked about acquiring sperm for the purpose of inseminating the patient's fiancée. The father went so far as to masturbate the son to obtain sperm for a semen analysis, which showed that the son had a normal sperm count and normal mobility.

In another case, an eighteen-year-old only child sustained a severe head injury and was declared brain dead within forty-eight hours. The father, who had undergone a vasectomy but who now wished to preserve the family lineage, requested the son's sperm be preserved for future insemination of a surrogate. The electroejaculation was performed, but it turned out that the patient was infertile and no viable sperm were obtained. The father decided to reverse his vasectomy in order to have a child via surrogate motherhood.

In Milwaukee, a man froze sperm before undergoing chemotherapy for cancer that would render him sterile. His plan was to have a child once his health was restored, but the cancer treatment was unsuccessful. After his death, his parents got a call from the hospital, asking what they wanted done with their dead son's sperm.

The couple began searching for a woman who would bear their son's child and let them act as grandparents. They found such a woman, but two insemination attempts failed. To increase their odds, they subsequently established a post office box (BABY, P.O. Box 10936 . . .) and began to offer the sperm to single women and married couples to create as many children as possible.

But is it really feasible that their son would have wanted to spread his seed across Wisconsin?

In a similar case, a young woman named Julie Garber, about to undergo cancer treatment, got an IVF clinic to fertilize her eggs with a donor's sperm and freeze the twelve resulting embryos. Her plan: to bear and raise her children after her recovery.

She died, though, and her parents laid claim to the embryos. "She passed away six months before, and I was in the same room, six feet away from her living cells," says Jean Garber. "There was something very special."

The Garbers began interviewing the eighty women who applied for the $15,000 job as a surrogate mother. One applicant said, "I enjoy so much being a mother I want to give Julie that chance."

Give Julie the chance to be a *mother*? Julie was *dead*.

The Garbers found an appropriate candidate, a postal carrier, and transferred the embryos into her womb. They arranged to appear on the *Today Show* with her—a family unit for the 1990s. When she miscarried the day before the show, they didn't divulge that fact to the producers. Julie's father, Howard, an ophthamologist turned cable television host, said they wanted to let other women know about their chance to become mothers from the grave.

The new technologies are broadening the divide between the reproductive haves and have-nots. At the same time that the wealthy Howard Garber was attempting to implant his dead daughter's embryo in a surrogate, he was lobbying against the poor having children.

"Thoughtless, careless and irresponsible people are perpetuating their cycle of poverty by having children they can't afford," he told the *Orange County Reporter*. "If you can't feed 'em, don't breed 'em."

Reproductive technologies are being marketed with short-term gains in mind, without any analysis of the long-term psychological or social impacts. A wife offered the opportunity to save sperm from a beloved husband who has just died may be grateful for the chance to feel as if she is keeping him alive just a little longer. But the net effect may be simply to prolong the grieving process. What happens to the wife who remarries and decides to have a child? Will she feel guilty if she doesn't use sperm from her dead husband?

Already, a New York man has requested that he be given the embryo he and his wife froze before her death—so that his *new* wife could carry it to term. In Tennessee, a man wanted his divorced wife's embryo implanted in his second wife. To my mind, it's bad enough that wife number two has to live in the same house the first wife did and sometimes wear some of her clothes. But to carry her embryo as well?

In Italy, baby Elisabetta was born in December 1994. Her mother had died in a car crash, leaving a frozen embryo that her paternal aunt carried to term. The aunt and her husband are raising her in the same home with her biological father. So the child's birth mother is also her aunt, while her genetic father is also her uncle.

As a result of Elisabetta's case, the Italian Parliament is considering a law banning surrogate motherhood and postmortem pregnancies. Says a representative of the Guild of Catholic Doctors, "Although we would say these embryos are life, doctors are not bound to go to extraordinary lengths to find a host to carry them."

The problem is mushrooming as a result of a growing trend where women store embryos before cancer therapy that will make them sterile. At Bourn Hall, Robert Edwards's former facility, the new director, Peter Brinsden, says he has frozen embryos from thirty such women. Now four British widowers are seeking surrogates to carry babies from their dead wives' frozen embryos.

A Houston company, Cryogenic Solutions, is capitalizing on the baby boomers' tendency to not want to close off any option. It offers women undergoing abortions the chance to freeze their fetuses, with the thought that, if technology is later developed to (in the company's words) "reanimate" the fetus, the woman can undo her abortion decision and raise that same child at a more convenient time. The service, "pregnancy suspension," costs $356. The company, which trades on the NASDAQ Stock Exchange, is in the process of patenting the procedure.

Geneticist Angus Clarke is frustrated with the fact that reproductive and genetic technologies tend "to be adopted as a matter of course once they become technically feasible, without a careful assessment of the ethical issues involved." He faults the medical profession for claiming that "the ethical questions are faced, and answered, by the families who consult us: it is their decision and we wash our hands of any responsibility."

At the Midwest hospital with the six comatose men, I tried

to discourage the doctor from electroejaculating his charges. If men do want to give their wives such a right, they can fill out a donation card, as they do to donate organs. When I later mentioned this in my law school class, several female students got their husbands to give them power of attorney over their sperm.

Because I believed that collectinsg sperm after death was akin to rape, I urged legislators not to allow it. New York state senator Roy Goodman introduced a bill now pending in the New York legislature, banning posthumous sperm collection unless the man had previously consented. Already, such a law is in effect in England and is being widely enforced after a case in which a woman, Diane Blood, was impregnated with sperm from her comatose husband.

The incentives for disturbing the comatose—or the dead—are high. Doctors are always looking for new technical challenges, and the financial opportunities for wives who create children after their spouse's death can be quite compelling. After her husband's death, Nancy Hart used his previously frozen sperm to create daughter Judith, then sued the Social Security Administration for death benefits for the little girl. The media put pressure on the agency (a typical article read, " 'Miracle' Baby Denied Death Benefits"), and SSA head Shirley Chater ultimately agreed to pay the girl $700 a month.

Legally, the SSA was not obligated. The law requires that children be compensated to replace support that their father gave them while he was alive, but this father had *never* supported Judith. Yet the fascination with reproductive technologies enchanted even the stodgy Social Security Administration into paying for their use.

Nancy Hart subsequently entered law school, with an express interest in the continuing evolution of reproductive technologies and their social implications. "Figuring out the inheritance rights of a cloned child will make our case seem simple," she says.

In time, perhaps she too will be fielding calls like the one I got from a lawyer in the middle of the night. A man had been fatally victimized by the police, and his wife had convinced the coroner to save some sperm. Now her lawyer was calling me to see whether, with sperm on hand, he now had *two* wrongful death claims against the cops, a prospective child's as well as the wife's. I explained, much to his disappointment, that unless the sperm was turned immediately into progeny, he would not be able to double his financial recovery.

15

The Clone Rangers

◞

The silver-gray UFO stands beckoning, its flying saucerish surface comfortingly smooth and curved, almost maternal. When I ascend the stairs to the hollow center, my voice echoes loudly through the chamber. There are no switches, computers, or other paraphernalia common to garden-variety spacecraft, so the science that runs this machine must be highly sophisticated. The room is empty except for two small swivel chairs, designed to accommodate aliens about four feet tall, just the type who Claude Vorilhon says abducted him back in 1973.

I am in a remote farming village in Valcourt, Canada, in the summer of 1998, to interview Claude, who now goes by the name Rael. Since 1973, he and his 40,000 followers— the Raelians—have been preaching the gospel according to the Elohim, the aliens. One of the central tenets of their religion is "Thou shalt clone."

Within a month of the birth announcement of Dolly the sheep, Rael had formed a company in the Bahamas, Valiant Ventures, and started Clonaid, a project to clone humans. One hundred couples signed up almost immediately. The first were people who were unable to conceive children together, some infertile, some gay. They were followed by parents who wanted to clone their dying children. With a price tag of $200,000 per cloning attempt, Rael soon had a promised pot of $20 million, enough to start hiring scientists to begin developing the procedure.

In addition to Clonaid, the company will charge $50,000 for a service called Insuraclone, to store cells from a living child in order to create its clone if the child is later lost to an incurable disease or accident. "If one of my children was dying, I would clone him. I have lost a child at six month's pregnancy and that was terrible," says Clonaid's scientific director, Brigitte Bosselier.

Another service, which the company hopes to provide soon, is called Clonapet, aimed at affluent individuals who wish to see their dead pet be "brought back to life." The market for such a service may be even more lucrative than cloning children. A wealthy couple recently gave $2.3 million to Texas A&M University to clone their dog, Missy.

Brigitte, who earned a doctorate in France, and another Ph.D. in biomolecular chemistry from the University of Houston, was fired from her research job in France when shortly after Dolly's birth, she first proposed human cloning. Her ex-husband is trying to have her declared an unfit mother. Yet she will not be deterred in her quest. "Parents have a right to have a baby who will have the genetic code of one of them," she says. "It is now common to see the dead parent father a baby through the process of frozen sperm implantation. Imagine the joy of a widow raising a child looking like her dead husband."

Rael says the aliens who contacted him clone only the most worthy people, but he is more practical. At least to start, he is willing to take anyone who can pay the $200,000.

Since they can't accommodate all one hundred patients at once, I ask how they will choose who will go first. "The person will have to agree to interviews by the media," says Brigitte.

"And maybe those who are willing to pay even more will go first," Rael adds.

Rael has been preaching cloning since 1973, but since the science initially was not available, he had to attract members by other means. Some young men and women were recruited by the lure of sex—at rave clubs, sex parties, and through ads in the personal pages. Segments of some Raelian meetings included sex, a dramatic variation of the part of the Catholic mass where you shake hands with your neighbor. And on Sundays at 11:00 A.M. they gather to contact the aliens. One British journalist who participated later pointed out that "weak magnetic fields can be used to induce exactly the same sensations experienced by abductees—a sense of tense, odd sexual fantasies and the immovable conviction they have just become interesting."

• • •

I had come to Dorval Airport in Montreal, where I was to make contact with the Raelians and then be driven two hours into the Quebec countryside to meet Rael himself. One of my colleagues had warned me that the group sounded like the cult in the Jonestown massacre. Another colleague had pulled out Rolph Ehrlich's phone number, the ex-military man now in the personal protection business, and asked me to call after my interview with the Raelians to assure her I was okay.

I was to be met at the airport by Lear. "*Lear* is *Rael* spelled backward," he told me over the phone. "I am Rael's assistant." I was to look for a man six feet tall with long blond hair wearing a large medallion with the Raelian symbol on it.

I had no idea what the Raelian symbol would look like. It used to be a cross between the Star of David and a swastika, but since the Raelians are negotiating to build an embassy in Israel, they have redesigned.

Lear was wearing a loose-knit gun-metal-blue top and white pants with blue and gray stripes. He is a bishop in the Raelian church, one of seventeen who serve directly below Rael. His territory is North and South America. Another bishop has Asia, still another Africa. He has been a Raelian for twenty years, since he was fourteen.

"The other religions are for life as it was thousands of years ago," Lear told me. "We need a religion that is in harmony with life as it is now. We need a religion that is in harmony with science."

The Raelians believe that extraterrestrials initially created all life on Earth in scientific laboratories using cloning and occasionally inseminating women who descended from the original group. Rael believes that on December 25, 1945, extraterrestrials took Marie Colette Vorilhon into their UFO to inseminate her in order to produce him, their final prophet, who was born in 1946. When he was twenty-seven years old, he was contacted by the aliens while he was hiking in a dormant French volcano.

They changed his name to Rael and told him to meet them the next day with his Bible. They met a half a dozen times on successive mornings to comment on the most significant parts of the Bible, revealing that *Elohim* was originally mistranslated in the Bible as "God," but it really means "those who came from the sky."

The Raelian calendar starts on August 6, 1945, the day the bomb was dropped, since that was the day, says Rael, when humans began the scientific odyssey that would lead them toward cloning. Hermann Muller's work on genetic mutations caused by radiation won a Nobel prize, and the Department of Energy began investigating the genetic damage done to survivors of Hiroshima and Nagasaki. After bombs had become less fashionable, in 1986, the DOE reconceived its labs as a place to investigate disease mutations in genes—and the Human Genome Project was born.

Rael says that the Elohim who came to Earth, like the pilgrims who settled the United States and Australia, were outcasts on their own planet. "They had the same level of civilization that we have now, with cloning and travel in space," says Rael. "But public opinion and ethical opinion were against cloning. So the scientists said, 'Let's go someplace else and do it.' "

"It's hard to be a pioneer," says Rael. "Ninety-nine percent of the Catholics today would not have followed Jesus Christ because it was new and not politically correct. It was the same with Buddha or Mohammed."

I try to conduct my interview with Rael and scientific director Brigitte Bosselier in the serious manner befitting a talk with a prophet. The setting makes it difficult, though. We are in the UFO Café in UFOland, a Raelian theme park, a miniature Disneyland. We are sitting at a table with a plastic tablecloth with a design of pumpkins, artichokes, and corn. The vaguely

Halloween motif of the tablecloth captures the spirit of the meeting. Brigitte Bosselier is dressed like Cleopatra. Rael is dressed in a white Elvis jumpsuit.

We are just a few feet from the glass souvenir cases where you can buy alien key rings, Japanese baby socks decorated with flying saucers, laser sunglasses, robot clocks, a puzzle depicting the solar system, or any of the many books that Rael has authored. Rael's *Sensual Meditation* book and video explain how to enhance body pleasure and reject Judeo-Christian guilt.

Yet as I explore the exhibits in UFOland, including the world's largest replica of DNA (twenty-six feet tall) and a room that is like the inside of a human cell, I am struck by how everything, except the flying saucer itself, looks like part of the educational material from the Human Genome Project.

• • •

Rael says clones will be better off than children born through sex, because they will be less of a disappointment to their parents. "Sometimes fathers are disappointed in their sons," says Rael, but there will be less parental regret if the son is identical to the father.

It will be good for the child, too, says Rael. "Usually children are happy to look like a father or mother," he says. "If the parents want to send him to the same school, why not? If the father is a brilliant scientist or painter, why shouldn't the child be, too?"

"I've met the parents," says Brigitte. "They are asking questions, talking to psychologists. Some have come with their IVF doctors, saying they've tried everything to have a child. They're not going to tell the baby he or she was cloned."

Withholding this information seems problematic, however,

especially if she is choosing the first clones based on their willingness to go public. And it didn't work with sperm donation. When families tried to hide their child's unique beginnings, the truth often came out at inappropriate times. The father would blurt out in anger, "No wonder you're a problem. You're not my child anyway." Or the parents would divorce, and the mother would try to stop her ex from seeing the child on the grounds he wasn't the real dad.

Now there are books, similar to the adoption books for kids, to explain to the children that they were created with the help of a sperm donor or surrogate. Would we now need books that began, "Mommy and Daddy wanted a child so much, they created you as a clone?"

In the future, Rael says, cloning will be a form of eternal life, where there will only be one clone of an individual at a time, and that clone will be implanted with the memory of the original person.

"Except," says Lear, "stupid people won't be cloned. Imagine how bad it would be to be stupid for eternity."

I raise my concerns against cloning. Physical risks, like the deformed cloned frogs or the cloned sheep that died shortly after birth. And the social risks of a hand-me-down genome. If we cloned Michael Jordan and the original died at age forty of an inheritable cancer, his clone would be uninsurable.

"Gene therapy would fix it," says Rael.

What will happen until gene therapy is perfected? After all, the gene for sickle-cell anemia was discovered in 1949, and there is still no cure.

"Then why not charge him higher insurance rates?" asks Rael. "Why does society have to bear the burden? It's a choice between the lottery and reality."

Concern about risks was leading some countries to ban

cloning. "The positive aspect of the retarded planet we live on is that the different countries have different laws," says Rael. That way, human cloning will always be permissible somewhere.

"I can't wait to be on *Larry King Live* with a beautiful couple and showing a blond little boy, asking people, 'Why are you against cloning?' " says Rael.

One stumbling block to Rael's desire to send in the clones is that the Roslin Institute, which cloned Dolly, has a patent on the cloning technology—and refuses to license it to the Raelians.

"The patent is not valid," asserts Rael. "It is as if the first person to patent a car had the rights to all future cars."

Brigitte believes they can clone without infringing the Roslin patent. "We can do it better," she says, pointing out that a Japanese group has come up with a different, more successful approach than Roslin's. She envisions discovering the chemical in the egg that causes the adult cell to reprogram itself, and using that chemical directly, rather than using the Roslin approach of injecting the adult's DNA directly into a donor egg.

"What we'd like to do is build our own universities and do our own research," says Brigitte, as people of other religions—the Jews and the Catholics—have done.

Rael points out that other religions also revere science. "The Bible says, 'Man is nothing without science,' and the Koran says, 'The blood of a scientist is worth more than the blood of a prophet.' "

As Rael is speaking about science, I have a flashback to the previous Sunday morning, when I was listening to the speakers at Northwestern University's Summer Biotechnology Institute in the ballroom of Chicago's Intercontinental Hotel. Each speaker got up and provided a dazzling series of scientific break-

throughs—gene therapy, bionic organs. They were total enthu-
siasts for science. None mentioned any risks.

• • •

Two days after the *New York Times* announced the birth of the
cloned sheep Dolly, Randolfe Wicker, the silver-haired, blue-
eyed owner of Uplift, Incorporated, an art deco lamp store in
Greenwich Village, incorporated Clones Rights Action Center.
He began organizing Clones Rights United Front, leading
demonstrations toting a placard with a picture of a sheep kneel-
ing on a cloud, forelegs outstretched, and the slogan DOLLY
LAMA—OUR NEW SPIRITUAL LEADER. He testified in the New
York Senate against a proposed bill to ban cloning.

In describing his pro-cloning activities in his 1997 Christmas letter, Randy said, "Yes darlings . . . I feel more and more
pregnant these days."

His Christmas letter closes with "Clone Today! Here
Tomorrow!"

I first learned of Wicker not long ago by visiting his Web
site, www.gaytoday.badpuppy.com., but Randy is a longtime
activist. He was a Vietnam protester, selling more than 2 million antiwar buttons out of a New York button shop, and lobbying to legalize marijuana. When I visit Randy, he gives me
one of his new buttons: CLONING: REPRODUCTION WITHOUT
COMPROMISE.

"My decision to clone myself should not be the government's
business, or Cardinal O'Connor's, any more than a woman's
decision to have an abortion is," he says. "Cloning is highly significant. It's part of the reproductive rights of every human
being."

Ann Northrop, columnist with a New York gay newspaper,

LGNY, agrees, saying cloning appeals to gays because "in a time when we're afraid that discovery of a genetic basis (for homosexuality) would lead to people aborting us, cloning would be a way of surviving."

"My triumph will come when my genotype lives on and I can say, 'Too bad, Mr. Death,' " says Randy. " 'I got another go-round. Let's see what I can do this time out of the box.'

"No longer does death have to mean total loss, accepting the obliteration of 'going quietly into that good night.' "

Randy had considered having a child with a surrogate mother. "I'd get enough money and go somewhere like Mexico or some third world country and pay a woman to bear my child. But the problem with that is, half of the genes are going to belong to this woman."

He told a *New York Press* reporter, "With this cloning business, I'm really working out some unfinished personal business. I was an only child who cried himself to sleep at night. I always wanted lots of brothers and sisters and to have lots of children, but, as it turns out, fate made me gay. So I think that's really my motivation for wanting to be cloned."

Wicker's mother told him that, at age sixty, he was "too old to be changing diapers." So he has decided he would like his clone adopted by a liberal heterosexual couple. "I'd like to be that special uncle in the family that everyone always wants.

"I wonder," he asks, "will I be jealous if my [cloned] son is heterosexual and has a child with a super girl?"

Randy and I settle into a tearoom, just down Christopher Street from his lamp shop.

"If you are against cloning, how do you feel about naming your child Junior?" he asks me.

I admit that it could create some of the same psychological problems for the child.

"I'm against naming the child Junior," he admits.

Randy confesses that he was a junior: Charles Gervin Hayden, Jr. Then, in the 1960s, he came out of the closet—and onto radio talk shows. It embarrassed his dad. So he remained Junior by day, but at night he took on a whole other identity, with a newly made-up name: Randolfe Wicker.

But even Wicker has his limits. "I wouldn't clone a dead child," he says. "I'm afraid that replacing a lost child through cloning would not necessarily be a good idea. Given the power structure between parent and child and the intense sense of loss and desire 'to replace' the one lost, I think this may be one of those few circumstances in which a preexisting identity being forced upon the later-born twin could actually impair his/her right to uniqueness and identity."

He also admits that cloning might not work out the way people expect. Wicker led a rally to clone Princess Diana. On September 19, 1997, he distributed nearly 5,000 commemorative badges reading: CLONE DIANA! ONE GOOD LIFETIME DESERVES ANOTHER!

"But Diana had an eating disorder," he tells me. "You might end up with a three-hundred-pound, later-born twin of Diana, which is not what you were thinking of when you cloned her."

16

A Clone-Free Zone?

❧

Within a week of Dolly's birth, an in vitro fertilization clinic began asking patients if they would be interested in using cloning if it became available. Many said yes. A woman in the United Kingdom sought to clone her deceased father. A female separatist group applauded the development of cloning, since it would make men unnecessary. And President Clinton asked his newly formed National Bioethics Advisory Commission to make recommendations about human cloning. The commission, in turn, called me, asking that I drop everything else to provide a legal opinion to the president.

I thought of this as the Bill Gates problem. If a wealthy individual like Gates wanted to clone himself—perhaps creating Bill Gates 5.0, 5.1, and 6.0—would any existing law stop him?

I asked embryologist Don Wolf at the Oregon Primate Center how much he thought it would cost for the equipment and personnel necessary to clone a human. Wolf had just cloned two rhesus monkeys using embryonic cells, rather than the adult cells used for Dolly. In addition to his research with chimps, he worked with patients in a human IVF clinic.

"About a million dollars," Wolf said. He himself wasn't interested in cloning humans. He was too busy. The National Institutes of Health were virtually throwing money at him since the announcement of his breakthrough cloning primates. They wanted identical hordes of monkeys for new drug research.

As I reviewed existing laws, I saw that there was nothing there to stop the truly rich from cloning. Certainly not the million dollars. Nine states banned experimentation on embryos, but those laws wouldn't apply. The experimental cloning procedure—injecting an adult's DNA—occurred with an *egg*. Once an embryo developed, nothing novel would be done. It would be inserted in a woman's womb using the same technique as in standard in vitro fertilization.

"I have to comply with a dozen federal regulations before I even sneeze at a mouse," says David Cox, a Stanford geneticist who served on the ELSI Working Group and is now a member of the National Bioethics Advisory Committee. "But I was shocked at how few laws protect people in human experimentation."

The acting director of the Food and Drug Administration weighed in with his view that his agency had the power to regulate cloning. But if that was true, why hadn't the FDA regulated in vitro fertilization or any of its variants? Like cloning, some IVF techniques—such as ICSI—had both a low success rate and potential genetic risk to the resulting children. And even if the FDA was able to assert authority, it only had the power to assess safety and efficacy. It had no right or expertise to decide whether cloning should be banned on moral grounds.

Nor were there any clear laws to prevent Bill Gates from being cloned against his will. What if Gates's barber used DNA from hair follicles to create a Gates clone—and then sued Gates for child support? Under current law, people have few legal rights to their body tissue and genes once these materials leave their body.

In the *Moore* case—the genetics case in which I'd filed a brief—John Moore's doctor had used Moore's unique tissue to develop a cell line worth millions. The California Supreme Court found that Moore had no right to a share of the proceeds, which could lead to a snip-and-run industry of cloning from stolen bits of celebrity hair. (Already, Nobel laureate Kary Mullis is marketing jewelry with celebrity DNA in it—why not a human replica?)

Even if Gates *intended* to clone himself, to hold power in his company or create a worthy heir, he might not be recognized by law as the legal parent. In some states, the legal parents would be Gates's parents and the replicant would be his brother. In two other states—North Dakota and Utah—if the twin were gestated by a surrogate mother, the child would be considered the legal offspring of the surrogate and her husband, even though she had no genetic connection. Gates would be a legal stranger to the child.

And what would Gates actually get if he cloned himself? Some geneticists claim that virtually all traits are inherited. Along those lines, psychiatrist David Reiss of George Washington University has declared, "The cold war is over in the nature-versus-nurture debate." But though identical twins reared apart show surprising similarities, they also have many differences. A cloned twin, raised later in a different historical period, would be even more different.

Of 277 attempts in the sheep cloning experiment, only one—Dolly—survived. Turning on the slumbering genes may activate hidden mutations.

Ian Wilmut, Dolly's creator, responded to the announcement

that Dr. Richard Seed intended to clone human beings within the next two years by stating, "Let me remind you that one-fourth of the lambs born in our experiment died within days of birth. Seed is suggesting that a number of humans would be born but others would die because they didn't properly develop. That is totally irresponsible."

Scientists do not fully understand the cellular aging process and, consequently, do not know what "age" or "genetic clock" Dolly inherited. On a cellular level, when the report of her existence was published in *Nature*, was she a normal seven-month-old lamb, or was she six years old (the age of the mammary donor cell)? There is speculation that Dolly's cells most likely are set to the genetic clock of the nucleus donor and therefore are comparable to those of her six-year-old progenitor, which could psychologically lead people to view cloned animals and humans as short-lived, disposable copies.

Keith Campbell, one of Wilmut's collaborators, said that we would never know if the cloned sheep would die earlier than others because almost all sheep become lamb chops long before succumbing to a "natural death."

Even if cloning posed no physical risks, the emotional impact could be devastating. If a cloned person's genetic progenitor is a famous musician or athlete, parents may exert an improper amount of coercion to get the child to develop those talents. True, the same thing may happen—to a lesser degree—now, but the cloning scenario is more problematic. A parent might force a naturally conceived child to practice the cello hours on end but will probably give up eventually if the child seems uninterested or tone deaf. More fervent attempts to develop the child's musical ability will occur if the parents chose (or even paid for) nucleic material from Yo-Yo Ma. And pity the poor child who is the clone of Michael Jordan. If he breaks his kneecap at age ten, will his parents consider him worthless? Will he consider himself a failure? What would that mean for the clone of Jordan's coach, Phil Jackson?

In attempting to cull out from the resulting child the favored traits of the loved one or celebrity who has been cloned, the social parents will probably limit the environmental stimuli that the child is exposed to. The cellist's clone might not be allowed to play soccer or just hang out with other children. The clone of a dead child might not be exposed to food or experiences that the first child had rejected. The resulting clone might live in a type of "genetic bondage" with improper constraints on his or her freedom.

Should cloning be allowed? We have drawn the line on some medical developments. Even though it is possible to transplant eggs from aborted fetuses into infertile adult women, we have chosen not to do so, in part because of the psychological impact on the resulting child when she learns that her mother died (indeed never lived) before she was conceived.

We don't have a system for debating the introduction of new technologies in the United States, but in Canada, the multidisciplinary Royal Commission on Reproductive and Genetic Technologies was formed. After two years spent assessing Canada's cultural values in a range of ways—from anthropological studies to public responses recorded on its toll-free number—the commission determined that Canadian cultural values disapproved of the objectification and commodification of people. Consequently, it recommended bans on cloning, as well as on paid surrogate motherhood and several other genetic and reproductive practices.

In the United States, though, we don't appear to have the

shared cultural values that might give rise to such a ban. Our attitude seems to be "Show me the money, and the technology will be available."

As I wrote my 113-page report on cloning for the National Bioethics Advisory Commission, in which I urged a ban on human cloning, I was struck by the fact that my past was coming back to haunt me. I had helped create the legal precedents that gave couples the right to use reproductive and genetic technologies. Now John Robertson, the lawyer who had replaced me on the American Fertility Society Ethics Committee, was testifying before the National Bioethics Advisory Commission that reproductive freedom included the right to create a child through cloning. The fact that cloning was so risky and so unlikely to succeed made no difference to Charles Strom, a member of the American Society of Human Genetics Ethics Committee. "I don't think that is an impediment," he told a reporter, pointing to the low initial success rate of IVF itself.

I thought about the testimony that Joseph Schulman, head of the for-profit Genetics and IVF Institute, had given before a National Academy of Sciences committee I had served on. "Don't regulate genetic technologies," he said, "because it will slow down their development. The computer industry developed quickly because anybody could tinker in their garage."

Yet with genetics and reproductive technologies we are tinkering with future people.

President Clinton issued an executive order banning cloning using federal funds. He also urged Congress to pass a law banning cloning with private funds. Harold Varmus, the new head of the National Institutes of Health, spoke out against the ban.

Three states—California, Michigan, and Rhode Island—adopted laws banning human cloning, but those laws are rapidly being outpaced by the technologies themselves. The California law prohibits injecting adult DNA into a human egg. But embryologist Neal First at the University of Wisconsin recently demonstrated that *cow* eggs could serve as a universal incubator for clones of other mammals. Human eggs—which cost $2,500 each—may no longer be necessary. Neal First got his cow eggs free from the slaughterhouse.

And now an infertile California man wants to challenge his state's prohibition on cloning on the grounds of reproductive freedom. But cloning is vastly different from normal reproduction or even the variations on IVF that the ACLU and I convinced a federal court to constitutionally protect. In even the most high-tech reproductive technologies available, a mix of genes occurs to create an individual with a genotype that has never before existed on Earth. In the case of twins, two such individuals are created. Their futures are open, and the distinction between themselves and their parents is acknowledged. In the case of cloning, however, the genotype has already existed. Even though it is clear that the individual will develop into a person with different traits because of different social, environmental, and generational influences, there is evidence that the fact that he or she has a genotype that already existed will affect how the resulting clone is treated by himself, his family, and social institutions.

In that sense, cloning is sufficiently distinct from traditional reproduction or alternative reproduction. It is not a process of genetic mix but of genetic duplication. It is not reproduction but a sort of recycling, where a single individual's genome is made into someone else. "Novelty is not progress," Boston University law professor George Annas told a congressional committee. "This change in kind in the fundamental way in which humans can 'reproduce' represents such a challenge to human dignity and the potential devaluation of human life (even comparing the 'original' to the 'copy' in terms of which is to be more valued) that even the search for an analogy has come up empty-handed."

Gilbert Meilaender, testifying before the National Bioethics Advisory Commission, pointed out the social importance of children's genetic independence from their parents: "They replicate neither their father nor their mother. That is a reminder of the independence that we must eventually grant to them and for which it is our duty to prepare them."

The risk here is of hubris, of abuse of power. Cloning, says lawyer Francis Pizzulli, represents the potential for "abuses of the power to control another person's destiny—both psychological and physical—of an unprecedented order."

Perhaps the best analogy to cloning is incest. Reproductive freedom is threatened as much by a ban on incest as by a ban on cloning. The harms are equally speculative. Yes, incest creates certain potential physical risks to the offspring, due to the potential for lethal recessive disorders to occur. But no one seriously thinks that this physical risk is the reason we ban incest. A father and daughter could avoid that risk by using contraception or agreeing to have prenatal diagnosis and abort affected fetuses. There might even be instances in which, because of their personalities, there is no psychological harm to either party. Yet we ban incest—despite the speculative nature of the harm—because it allows an exercise of excessive power of parents over children.

British philosopher Dr. Mary Midgley described the response to the news of mammalian cloning: "It is primarily a fantasy about power. What is disturbing doesn't seem to be the prospect of these things actually happening, but the kind of excitement which the power fantasies generate, not just at the street level, but perhaps also among the kind of people who control funds for scientific research."

There seems to me to be a world of difference between reproductive technologies (in vitro fertilization, egg donation, sperm donation, or surrogate motherhood), which allow couples to make up for a missing ingredient in the normal reproductive process, and the technologies now being proposed to let dead men beget children, to reanimate dead fetuses, and to create children with only one genetic parent. The former techniques meet existing needs; the latter create needs and try to shoehorn them into the existing category of women's reproductive choice. Moreover, cloning hardly seems to be an enhancement of *women's* freedom. When it comes to suggestions of who we should clone, it is *men* who make the list—Michael Jordan, Albert Einstein, Bill Gates.

Before my trip to Dubai, I tried to imagine what position the Muslims there would take on cloning. I wanted to use parts of my federal legal opinion on human cloning in my Dubai presentation, yet much of it seemed irrelevant, even blasphemous, from an Arab point of view. Any right to clone in the United States would be based on an analogy to the right to abort, but abortion is illegal in Dubai. A right to clone in the United States is being asserted by gays like Randolfe Wicker, but homosexuality is a capital offense there. Well, what about my psychological arguments against cloning?—that it involves excessive parental power over children and fosters too much uniformity. That wasn't going to be a persuasive argument to the Muslims.

I looked down at the first sentence I had written for the

speech. "There are 270,000 sheep in Dubai, but none as famous as Dolly."

I tried to think of how I could expand that for another twenty pages.

Once in Dubai, I had the chance to present my views to an audience almost entirely of males. It was the first speech I had ever given where members of the audience wore sidearms. At one point, a religious leader rose and asked me an angry, fifteen-minute question in Arabic. I stood quietly at the front of the room, listening to the simultaneous translation of his hostile words. How could I, he was asking, even mention technologies such as egg and sperm donation or surrogate motherhood? Didn't I know that they violated long-standing values, dating back to Adam and Eve?

As he embarked on his lengthy Old Testament interpretation, I looked at the guns and considered responding with "I agree completely" to whatever he said.

Instead, I listened carefully, then tried to put his concern into a larger context. I mentioned that many Catholics, as well as many other individuals, were concerned with the impact of reproductive technologies on their values and on society.

In the Middle East, a woman can be sentenced to death for undergoing artificial insemination by donor. Yet by the time the meeting ended in Dubai, Muslim religious leaders and Middle Eastern in vitro doctors had come to an accord that it would be consistent with Islamic values to clone *men*—infertile, married men—as long as it was used within the marriage relationship.

It was then that I realized that somatic cell nuclear transfer, as the Dolly technique was called, was not so much a scientific technique to reproduce individuals, as a way to clone our values.

The Raelians would clone smart people, because that was what they valued. The Muslims would clone men.

Perhaps the main objection I had to cloning was it replicates everything that was troubling about reproductive technologies: excessive commercialization, reckless experimentation on women, procedures undertaken without consent, unmonitored physical and psychological risks. From my point of view, it was time to reverse the process. Cloning seemed to be the perfect opportunity to shift the burden of proof, to ask scientists to give a good reason rather than a false promise before they began the technique, to show why it was really necessary, and to design a system from the start to protect the participants.

My speech at Dubai was a mea culpa. I had helped make reproductive technologists invincible, and facing human cloning was like greeting Frankenstein's monster for the first time. The creation had gone amok. I needed to draw the line here to atone.

Afterward, Georges Kutukdjian, the UNESCO (the United Nations Educational, Scientific, and Cultural Organization) bioethics director, told me my speech had moved him. In December, UNESCO had adopted a "Universal Declaration of the Human Genome and Human Rights," which stated, "Practices which are contrary to human dignity, such as reproductive cloning of human beings, shall not be permitted." The response, by some scientists, was outrage and ridicule. But after Richard Seed announced he was going to begin human cloning, nineteen countries signed a ban against it. Juergen Ruettgers, the German research minister, said, "We simply cannot stand by and allow humans to be copied. That would be breaking through an ethical barrier that goes far beyond even the barrier of the atomic bomb."

The summer after Dolly's birth, gene therapy pioneer W. French Anderson and gene sequencer J. Craig Venter hosted the First International Congress on Mammalian Cloning. I spoke against human cloning, as did Ian Wilmut, Dolly's creator, and Alex Capron, a member of the National Bioethics Advisory Committee. Ron James, a researcher himself and the venture capitalist behind Wilmut's work, urged the researchers in attendance not to begin human cloning. Even though he could make untold millions if humans were cloned (since his company had rights to the patent on the mammalian cloning technique), he opposed it on eugenic grounds.

"My father put beer bottles in cases for a living," James said. "My mother cleaned houses. If cloning had been around, I would never have been allowed to be born."

A year later, in June 1998, at the *Second* World Congress on Mammalian Cloning, the tenor of the discussion had changed dramatically. Philosopher Gregory Pence spoke, whose book *Who's Afraid of Human Cloning?* advocates the procedure. So did Lee Silver, the Princeton biologist whose book *Remaking Eden* celebrates the science of genetics and reproductive technology, including human cloning.

"You used to be so smart," Silver had written to me, "I can't understand why you could take a position against cloning."

His fax to me mirrored a letter from Randolfe Wicker, head of Clones Rights United Front. "I am convinced that you will eventually be on our side in the cloning reproductive freedom debate," Wicker wrote. "I recognize an open inquiring mind when I see one."

It was a third panelist, though, who most signaled the change in scientists' views. For their 1998 meeting, the conference conveners had chosen Brigitte Bosselier, the Raelian scientific director of Clonaid. The meeting provided the perfect opportunity to recruit scientists to enable her cloning venture.

"Dr. Seed is saying what everybody else is thinking quietly. It's true that, once we can clone a human, that's the first step towards eternal life," Bosselier said. "If we can reach some kind of eternal life then God disappears. Too many people are trusting their belief in God, but we are going beyond that. Science is our religion, and you can't stop science."

I thought about how three decades ago, Sophia J. Kleegman and Sherwin A. Kaufman, in *Infertility in Women*, observed that new reproductive arrangements are greeted initially with shock and must pass through several stages before they are accepted: "Any change in custom or practice in this emotionally charged area has always elicited a response from established custom and law of horrified negation at first; then negation without horror, then slow and gradual curiosity, study, evaluation, and finally a very slow but steady acceptance."

With artificial insemination, acceptance took decades; with in vitro fertilization, it took years. The attitude toward cloning shifted in a matter of months.

At the Mammalian Cloning Conference, Dr. Richard Seed, the reproductive entrepreneur I had met years ago in Kiel, took me aside. He gave me his card and told me that for $3.5 million he was willing to clone a person. I put the card in my purse. But as I walked away, I thought to myself, *He'll have to find a lawyer other than me to do the legal work.*

AFTERWORD

In November 1998, a new boutique opened in a trendy shopping area in Pasadena, California. The shop was staffed by two bespectacled intellectuals, Tran T. Kim-Trang and Karl S. Mihail, both wearing white laboratory jackets. Just two hours north of the Repository for Germinal Choice in Escondido (the sperm bank selling the seed of Nobel Laureates and top athletes), the boutique, Gene Genies Worldwide, offered "the key to the biotech revolution's ultimate consumer playground." It sold new genetic traits to people who wanted to modify their personalities and other characteristics.

The outside window of the boutique was stenciled with a whimsical outline of an Aladdin's lamp, but the inside was filled the vestiges of biotechnology—petri dishes, a ten-foot model of the ladderlike structure of DNA, books by Tran and Karl such as *Imitating Nature Today, Subsuming It Tomorrow*. Brochures highlighted traits that studies had shown to be genetic: creativity, conformity, extroversion, introversion, novelty seeking, addiction, criminality, and dozens more.

Hundreds of unsuspecting shoppers dropped by, some strolling in from the Johnny Rockets restaurant next door, or the AMC theaters across the plaza. Some were horrified, accusing Tran and Karl of

Nazism. But, more commonly, people pulled out their credit cards, ready to order new genetic traits for themselves and their children.

People initially requested one particular trait they wanted changed, but once they got into it, their shopping lists grew. Since Gene Genies offered people not only human genes but ones from animals and plants, one man surprised Karl by asking for the survivability of a cockroach.

The co-owners were thrilled at the success of their endeavor, particularly since none of the products they were advertising were actually available yet. Despite their lab coats, they were not scientists, but artists attempting to make a point, striving to serve as our moral conscience. "We're generating the future now in our art and giving people the chance to make decisions before the services actually become available," says Karl Mihail.

When I met Karl and Tran, I was struck by how much what they were trying to do paralleled what I try to do in my own work. I am trying to get people to think about where reproductive and genetic technologies are taking us—and to make some judgments about whether we really want to go there.

In the year since I wrote this book, scientists and doctors have continued to push the boundaries, with women and children suffering the consequences. Nkem Chukwu, a twenty-seven-year-old woman in Houston gave birth to octuplets. Chukwu had used a fertility drug, human gonadotropin. Just four months into the pregnancy, she was put on complete bedrest. She spent her last month in a special, tilted hospital bed to keep pressure off her cervix. The powerful anticlotting drugs she took to protect the fetuses made her nose bleed. She coughed blood. After the birth she hemorrhaged, losing twelve pints of blood (more than a person's whole supply). She nearly died. One of the babies died. The seven surviving children were subjected to all manner of medical interventions—catheters, electrodes, ventilators, sedatives, steroids, antibiotics. Three had eye surgery, two had intestinal surgery.

The McCaughey septuplets have had their own health problems.

Natalie and Alexis have difficulty eating. Surgery did not correct the problem and now they are tube-fed, sometimes vomiting uncontrollably. Nathan and Alexis have muscle disorders and cannot sit up or walk without assistance. Kenny had eye surgery.

In animal research, new problems have arisen. Although Dolly the cloned sheep is three years old, the telomeres on the end of her chromosomes look more like those of a nine year old, suggesting they started out at the age of her six-year-old progenitor. This led to a great headline in *The Washington Post*—"A Sheep in Lamb's Clothing"—and concerns from some scientists that she might prematurely age. In fact, about half of the sheep and cow clones have serious abnormalities—defects of the heart, lungs, or other organs—and some have died before birth. Others seem healthy at first, then die from problems such as immunological disorders.

Yet the proposals for cloning human beings are growing. Nathan Myhrvold, an executive at Microsoft, says that prohibiting human cloning is a form of "racism."

The march of "firsts" has continued in this field. Gaby Vernoff gave birth to a baby girl created with sperm collected four years earlier from the corpse of her husband Bruce. The $35,000 process, paid for by her in-laws, involved fertilizing Gaby's egg in vitro with a single sperm from Bruce.

Previously, other women conceived children after their husbands' deaths using sperm that had been collected while the men were still alive. Some of those children are now entering school, and the mothers have been surprised at the emotional impact of posthumous parenting on the children. "I thought that since she had never had a father she wouldn't miss him," one of the mothers told me. "But she asks about him constantly."

And while men are procreating after death, women are being enticed with the possibility of motherhood post menopause. Sure, that has been done by women as old as sixty-three, using eggs from a younger female donor. But now women are being offered the chance to have their *own* children later in life by freezing a piece of their ovaries when they are young.

The procedure was performed in 1999 on a thirty-year-old American ballerina, Margaret Lloyd-Hart, who had prematurely entered menopause after her ovaries were removed years earlier for medical reasons. Bits of ovarian tissue that had been frozen were reimplanted in her womb. Eighty such tiny pieces were stitched onto a small triangular piece of dissolvable foam, then sewed into the spot where her ovary had been. Amazingly, this caused her to menstruate and release an egg. Even though she has not yet tried to get pregnant, doctors are beginning to promote this technique to allow young and middle-aged women a chance to put parts of their ovaries in "suspended animation" for later use.

In England, a little-noticed birth was perhaps the most portentous. Louise Brown's younger sister, Natalie, also created through in vitro fertilization, had her own child the old-fashioned way. In 1999, seventeen-year-old Natalie gave birth to a daughter, Casey. This showed that test-tube babies (at least the women) were capable of reproduction. But even Robert Edwards cautions that you can't tell much from a sample size of one.

While the bells and whistles surrounding reproduction have grown— in the form of posthumous and postmenopausal pregnancies—in vitro fertilization remains the most far reaching and significant technology. Its true potential (for good and for ill) is still being realized. Two decades ago, biologist Clifford Grobstein observed that once the embryo is isolated in the petri dish we have seemingly limitless options for studying and transforming it. Doctors can undertake preimplantation genetic screening on the embryo, or even add new genes to the future child. Biotech companies can collect human embryos as the raw material for pharmaceutical products, and divorcing couples can go to court to fight for custody of their eight-cell offspring.

"My husband is leaving me for another woman," the voice on the other end of the phone declared. I looked at my clock. It was a little after seven in the morning. Why was this stranger calling me about her marital woes? Then she got to the punchline. "We have six embryos frozen and I want to make sure I can use them after the divorce."

Such calls are common to me. As many as twenty thousand embryos are the subject of disputes, says New York state senator Roy Goodman, who held legislative hearings on the subject. It's not just divorce that leads to sticky questions about the fate of frozen embryos, but death as well. A New York man has requested that he be given the embryo he and his wife froze before her death—so that his *new* wife could carry it to term. Such situations are mushrooming as a result of a growing trend where women store embryos before undergoing cancer therapy that will make them sterile. British infertility specialist Peter Brinsden says he has frozen embryos from thirty such women. Now four widowers are seeking surrogates to carry babies created from their dead wives' frozen embryos.

The embryo disputes continue to perplex judges. If embryos are considered people, they would have a legal right to be born. But abortion would then be considered a crime. Yet labeling embryos as property does not seem appropriate, either. What would that mean in a divorce case? Honey, if you get the stereo, I get the embryo.

Since there is no easy legal category for human embryos, courts have been stumped. In a New York case, a couple spent over $75,000 on in vitro fertilization and then divorced. The trial judge gave the five frozen embryos to the wife, but an appellate court allowed the husband to donate them for research.

In Tennessee, a trial court judge ruled that a couple's seven embryos were "children" and had a right to be born; he granted them to the wife. A French geneticist witness in the case said the parents had a moral duty to rescue these "tiny human beings" from the "concentration can." On appeal, an intermediate court decided to treat the embryos as property. The court awarded joint custody of the embryos to the husband and wife, leading them to wonder if they should each take a couple of embryos, like splitting up the furniture or CD collection. When the case reached the Tennessee Supreme Court, though, the justices ruled the embryos were neither people nor property. Instead, they ruled that the husband had a right not to procreate similar to a woman's abortion right. They gave the husband the okay to terminate the embryos to prevent his unwanted fatherhood.

After a nine-month legal battle, my early morning caller won custody of the embryos in her divorce. Her husband does not want to pay child support, so she had to agree that if she became pregnant she would never reveal the identity of the father. She is comfortable with that unusual resolution, but wonders how realistic the confidentiality provision is. Her friends know she and her husband froze embryos. Her husband will be visiting their existing children. If the baby looks just like her ex, won't the child eventually suspect the truth? And, if he does not, how will he feel to have a cipher in place of a daddy in the family tree?

Our society's lack of consensus on the legal, moral, and social status of human embryos wreaks havoc with reproductive and genetic technologies. Strong right-to-life lobbying continues to prevent the federal government from funding research involving embryos. This means that new embryo technologies are not covered by the federal regulations requiring informed consent and mandating advance review to assure safety and efficacy. And rights to use the technology are owned by the private companies that fund them, which often charge very high fees for their use.

The problems with the current approach were underscored when researchers at the University of Wisconsin announced in the past year that they could use human embryos to create potentially beneficial therapeutics. Researcher James Thomson and his colleagues used excess embryos donated for research by couples undergoing in vitro fertilization. The researchers allowed the embryos to divide for five days. Then they isolated from those embryos the most coveted of human cells: embryonic stem cells, primitive cells that can grow into every type of body tissue, including nerves, bones, and muscles. These cells could provide information about the biology of human development that will lead to many new medical discoveries. In the future, a doctor might be able to use stem cells to produce new heart cells and inject the cells into an ailing heart, which would then repair itself. Because a biotechnology company, Geron Corporation, financed Thomson's work, that company owns patent rights for treatments developed using Thomson's embryo stem cell technology. After the researchers' announcement, prices of Geron stock soared 31 percent.

Consequently, some IVF clinic directors realized that they possessed a potential treasure trove. One clinic head called me to ask if it would be all right if they sold unclaimed embryos in their program to a biotech company. (I said no.)

The debate over scientists' work with stem cells reveals serious loopholes in research oversight. To address these problems, Congress should extend federal regulations to cover all human research, whether supported by federal funds or not. For instance, an institutional review board should scrutinize the scientific merits of every study before the research begins, and safeguards should be in place to make sure that all the subjects are participating voluntarily and have been adequately informed about the nature of the research.

Guidelines are also needed to protect couples whose in vitro embryos may be used in research. Their participation may be coerced if they feel that they cannot readily refuse when their doctor (the key to their becoming parents) asks for their excess embryos for research purposes. Moreover, as embryos become valuable to biotech companies as sources of embryonic stem cells, doctors may increase the dose of fertility drugs to insure that multiple embryos are created—in effect, to manufacture more "excess" embryos. The fertility drugs given to a woman to produce more embryos can increase her risk of diabetes, blood clots, heart failure, and even death.

Currently, couples undergoing in vitro fertilization and embryo freezing are generally asked on the consent forms to check off what they want done with their excess embryos: donation to another couple, termination, or research. Such couples should be told *specifically* when their embryos will be used for stem-cell research. Some couples, who might be comfortable with allowing their embryos to be used in research to improve fertilization techniques, may be troubled by an experiment that turns what could have been their potential child into a kidney or cells that are for sale. The recipients of human tissue, too, should be told of its origins. Some people may not want treatment that uses cells derived from human embryos, just as some Jehovah's Witnesses will turn down treatments involving blood transfusions.

Clarifying the status of the embryo—and the rights of the couple—is imperative, as the march of technologies using the embryo continues. Recently, gene therapy researcher W. French Anderson proposed adding genes to fetuses in utero to correct medical problems. It is only a matter of time before we start genetically engineering embryos.

Since I wrote this book, I have crossed paths with an entirely new set of people who are employing reproductive and genetic technologies— artists. Tran Kim-Trang and Karl Mihail are part of a new art trend, using genes as a source of inspiration. But while the two Californians use their work to comment upon people's hubris in attempting to remake themselves and their children with genetic tools, other artists demonstrate how easy it would be to do so. Artists are actually creating genes, shaping the clay of life itself. A genetics laboratory at the Massachusetts Institute of Technology has an artist-in-residence, Joe Davis, who is putting secret messages into actual DNA. And Eduardo Kac, a professor at the Art Institute of Chicago, advises artists to become creators with a capital "C" and try to make new species.

All of biology is becoming a set of Tinkertoys. "When genes go from metaphor to material, a fundamental change occurs," says Eduardo. And it is not just artists who are interested in this potential. Parents-to-be want to give new genes to their offspring. Species boundaries are no longer relevant; in the future, people could be given the gene for the running speed of a cheetah or the night vision of a bat.

Spending time with transgenic artists, as they call themselves, I entered the current *Alice in Wonderland* world of biotechnology that is unregulated and available to anyone. At one Web site, I can order up genetic tests to see if I have genes associated with breast cancer or Alzheimer's disease, among others. A handheld genetic sequencer is available that would allow me—à la the movie *GATTACA*—to test the DNA of my dates. And, from a company called Clonetech, I can buy the building blocks of life to create entirely novel genes.

Eduardo, Tran, Karl, and I participated in a weeklong meeting in Austria in September 1999 in which geneticists, philosophers, lawyers, and others raised questions of what the proper use of reproductive and genetic technologies should be. For me, the meeting had a special significance. It was taking place just eighteen miles down the Danube from Mauthausen, the Austrian concentration camp where 200,000 Jews, homosexuals, Gypsies, and other "undesirables" were annihilated.

What use will our generation make of genes? And how will future generations judge us?

For the past five years, Eduardo has envisioned making a work of art by genetically engineering a living dog so that it would express the green fluorescent gene that is carried by the Pacific Northwest jellyfish. But the techniques that Eduardo proposes to use on the dog could very well be used on people. In August 1999, researchers in Atlanta removed a gene from a prairie vole, an affectionate, monogamous rodent that spends half its time cuddling. They transferred the gene to a closely related species, the mountain vole, which lives a promiscuous lifestyle. The recipient rodents did not become monogamous, but their brains developed to look like those of prairie voles and they became more cuddly and affectionate. Science writers began to speculate on the potential applications to humans. In the wedding of the future, would we not only promise to love, honor, cherish, and obey—but also to have a prairie vole gene implant?

The very boundaries of what is human are being changed by genetic technology. Yet hardly anyone in the public or the legislatures is paying attention. We might notice if a totalitarian government decided to innoculate all its citizens with a particular gene. But the change, the designing of children, is occurring much more subtly, as a result of individual choices through an open market. One couple offered $50,000 for an egg donor who is a smart, tall, Ivy League student. A man seeking to sell his sperm for $4,000 a vial established a Web site with his family tree claiming to trace his genes back to six Catholic saints and several European royal families. Thousands of couples turn to the Internet to find genetic parents for their future children. They view pictures of sperm and egg donors, listen to tapes of their voices, and review pages of descriptions of their physical features, their hobbies, their philosophies of life. Can purchasing single genes—rather than a person's whole packet—be far behind?

When I lectured in Eduardo's class at the Art Institute of Chicago, one student said to me, "Conservative Republicans might want to give children the genes for citizenship or eliminate the genes for homosexuality. But I am an artist. I would want to give my child a blue triangular head."

How is society going to judge such desires? Should certain genetic manipulations be allowed and others not? Is giving a child a gene protective against a deadly disease appropriate but manipulating genes to create a blue triangular head not? What about cases that fall in the middle—genes to prevent baldness or assure taller stature? And should we really exercise dominion over other species, changing their features at will?

The task for all of us in the coming years is similar to that of science fiction writers and transgenic artists. We need to think long and hard about what society would look life if reproductive and genetic technologies continue to multiply without adequate regulation.

Discussing these technologies, Lord Jacobovitz, the former Chief Rabbi of Britain, said, "The great challenge to mankind today is not only how to create, but to know when to stop creating. And when we celebrate a Sabbath to remind ourselves that G-d initially created this world, we celebrate not only his act of creation on the six days. We celebrate that he knew when to stop."

Lori Andrews
December 1999

ACKNOWLEDGMENTS

I want to thank the clever and creative research team, who helped me through this book—and played a major role in many of the debates described here: Fletcher Koch, Michelle Hibbert, Greg Kelson, Andrea Laiacona, and Laura Swibel. They are the first people I turn to when I get a call for a legal opinion about a groundbreaking, yet disturbing, new technology. They also can make me laugh or convince me to wear false tattoos. Thanks, too, to the newest members of the group—Julie Burger, Katie Mason, and Valerie Neymeyer.

I also am indebted to my son, Christopher, for thinking up the title of the book, asking provocative questions, and playing great music to write by. And to Bryant Garth, for lovingly putting up with me while I was writing. I have also learned much from close friends and colleagues who have influenced my thinking, and who have commented on drafts of the book—Lesa Andrews, Anita Bernstein, Eric Goodman, Hal Krent, Jim Lindgren, Timothy Murphy (who first coined the term "sperminator"), Dorothy Nelkin, Clements Ripley, Mark Rosin, Jim Stark, and Karen Uhlmann. A special debt of thanks is owed to Nanette Elster, the gifted lawyer who worked with me on virtually every issue in this book and who is now teaching doctors to grapple with the profound issues raised by biomedical technology.

And I couldn't have been luckier than to be in the hands of Amanda Urban, my agent, and William Patrick, my editor. Amanda is crystal clear in her vision and critiques. Bill, a novelist as well as an editor, was an invaluable and ingenious partner in this process, as he attempted to teach me to write with "less lawyer navy and more Armani black."

NOTE ON SOURCES

◗

I guess there are advantages to being a pack rat. Over the course of my twenty years of involvement in reproductive and genetic technologies, I have saved every trial transcript, court decision, proposed and enacted law, and academic article related to these issues. As it turns out, I have been fortunate in having a ringside seat for most of the major events in this field. I have maintained my copious notes as well as the draft and final minutes from each committee meeting, legislative hearing, and telephone conference call in which I participated. I have notes from each interview I conducted for the cases I handled, and for the books and the articles I wrote on these subjects. These include interviews done for my articles in *Parents* magazine, *Psychology Today*, *The New York Times* syndicate, *Vogue*, various American Bar Association publications, law reviews, and medical journals, as well as for my books, *Medical Genetics: A Legal Frontier, New Conceptions, and Between Strangers: Surrogate Mothers, Expectant Fathers, and Brave New Babies*. These materials provided the starting point for an attempt to describe my career—and the evolution of reproductive and genetic technologies—over the past twenty years. I also benefited enormously in the creation of this book from the excellent science journalism in recent years, particularly the articles by Robert Lee Hotz of the *Los Angeles Times*, Gina Kolata of *The New York Times*, Judy Peres of the *Chicago Tribune*, and Rick Weiss of *The Washington Post*. Scientists, physicians, attorneys, lawmakers, and the consumers touched by the technologies also gave generously of their time in the past two years to provide me with the latest information about the scientific, medical, social, and psychological import of the technologies being developed.

For those readers who would like further information about genetic and reproductive technologies, several books are particularly insightful and engaging: George Annas's *Some Choice: Law, Medicine, and the Market*, Robert Lee Hotz's *Designs on Life: Exploring the New Frontiers of Human Fertility*, Gina Kolata's *Clone: The Road to Dolly and the Path Ahead*, Mark Rothstein's edited volume, *Genetic Secrets: Protecting Privacy* and *Confidentiality in the Genetic Era*, Lee Silver's *Remaking Eden: Cloning and Beyond in a Brave New World*, and Daniel J. Kevles's and Leroy Hood's *Code of Codes: Scientific and Social Issues in the Human Genome Project*.

This book is, at its heart, a memoir. Other people's interpretation of the events I describe may differ. I welcome other contributions to understanding and addressing the reproductive and genetics industries. The goal of this book is to begin to shed light on scientific, medical, and policy decisions of major import that rarely come to the public's attention. I chose to recount my own experiences as a way to bring the larger issues into focus. If other authors later weigh in on the subject, this will only be to the public's interest.